The Anti-Landscape

Studies in Environmental Humanities
1

Studies in Environmental Humanities is a publishing forum of the Nordic Network for Interdisciplinary Environmental Studies. It includes both monographs and multi-author volumes that engage vital and timely questions concerning the intersections of culture, society and environment. The series editors welcome queries and manuscript submissions.

The Anti-Landscape

Edited by
David E. Nye and Sarah Elkind

Rodopi

Amsterdam - New York, NY 2014

Cover illustration: Walker Evans, *Erosion near Jackson, Mississippi,* 1937. Farm Security Administration – Office of War Information Photograph Collection, Library of Congress.

Cover design: Inge Baeten, Senti Media

The paper on which this book is printed meets the requirements of "ISO 9706:1994, Information and documentation - Paper for documents - Requirements for permanence".

ISBN: 978-90-420-3886-8
E-Book ISBN: 978-94-012-1169-7

Printed in The Netherlands

Acknowledgements

This publication is one result of the activities of the Nordic Interdisciplinary Environmental Studies research group (NIES), jointly founded in 2006 by Mark Luccarelli, Steven Hartman, and David E. Nye, from, respectively, the University of Oslo, and the Royal Swedish Institute of Technology, and the University of Southern Denmark. This book contains papers selected from the third NIES conference held in Odense, May 9-11, 2011, which was also the first meeting made possible through a networking grant from NordForsk. Sarah Elkind was able to help prepare and to participate in the conference as part of a year-long exchange arranged through the Danish American Fulbright Commission, which thereby also supported this endeavor. Finally, we thank the Center for American Studies and the Humanities Faculty of the University of Southern Denmark for local arrangements and practical assistance including the layout of this volume, by Elsebeth Jensen.

Table of Contents

Introduction

1

The Anti-Landscape

David E. Nye

The anti-landscape is a new term, but the phenomena of spaces that have become unlivable was already described in Marsh's *Man and Nature*. The anti-landscape is not necessarily the result of human interference, but may have natural causes. Humanly created anti-landscapes, both intended and accidental, emerged already in prehistory. Such sites have become more numerous and prominent as technologies have become more powerful, and awareness of them has increased since c. 1960. This essay defines the term, using J. B. Jackson's work on landscape as its point of departure. It closes by explaining the basic narrative that is latent in the idea of the anti-landscape, and that provides the organizing principle for this collection of essays.

Keywords: Anti-landscape; Aral Sea; J. B. Jackson; landscape; Lake Erie; narrative; pollution; recovery

1. Landscape and Anti-landscape

J. B. Jackson once defined landscape as, "A composition of man-made or man-modified spaces to serve as infrastructure or background for our collective existence." (Jackson, 1984, 8). In contrast, anti-landscapes are spaces that do not sustain life. They may arise from natural causes but are often the consequence of human action. Landscape is a shared creation. It is not something that human beings look at in the distance. Rather, landscapes are the humanly modified places where people live, including urban areas. Landscapes are never entirely natural, nor are they the opposite of culture. They are the infrastructure of collective existence. They may suffer damage but they can also be improved. They can be protected from erosion. Rotating crops can enhance fertility. Biodiversity can be strengthened through development of hedgerows and preservation of marshlands and forests. To survive over long periods, people must find ways to live within and to maintain their landscapes.

I sensed the need for the term "anti-landscape" in 2006, when I was thinking about how electrical blackouts transform a large city, making it potentially uninhabitable. (Nye, 2007, 72-84) The water pressure disappears, and anyone living above the third floor immediately loses access to water, and soon everyone does. Toilets cannot flush and bottled water supplies sell out. During a summer power failure, air conditioning dies, and many buildings, especially modernist glass towers, heat up to over 100 degrees and must be evacuated. Refrigeration fails, food spoils, and supplies rapidly dwindle. How many days can a city like New York survive without electricity? I explored these issues in *When the Lights Went Out* but realized that the concept of anti-landscape had much wider applications. (Nye, 2010, 130-132, 182, 203, 231-232.) At first I tried to use the word landscape plus various additional terms to describe this concept (such as "destabilized" and "non-sustainable"), yet these seemed inadequate. Then I recalled Jackson's fine definition of landscape, cited above, and realized that anti-landscape could be defined as its inversion.

These definitions are inspired by cultural geography and environmental history, but landscape is a complex term that has a somewhat different meaning from one discipline to another. In art, a landscape is a genre of painting that emerged in the last years of the Middle Ages and early Renaissance. The definition of a landscape as a two dimensional construction representing a three dimensional space can be seen as inherently a falsification, simplification, or idealization. It should not be mistaken for a mirror showing us what existed at the time the image was made. Scholars concerned with landscape paintings and photographs often take a deconstructionist turn and conclude that no image can ever be a window but only a distorted mirror that reveals not the past but artistic conventions, the artist's preoccupations, and the contemporary viewer's historically shaped assumptions. From this perspective, the meaning of each landscape painting changes over time, and it is not grounded in external reality but in an ongoing visual discourse. (DeLue, et al. 2008) From a historical and geographic perspective, however, paintings and photographs are only some of the many sources that can be used to reconstruct the past. While they cannot be read naively as "realistic" depictions of another time, neither can their historical context be ignored. Several papers in this volume engage the visual arts as a way to understand changing perceptions of landscape, and they do so with an awareness of history and also an understanding of what cannot be represented visually, such as toxins in the air, soil or water.

In literature, as in art history, the divorce of image from reality also seems unavoidable. The incompleteness and partiality of any representation seems patent. The landscape in a novel is almost never an exhaustive description of a place, but rather a series of references that suggest to the reader a setting, such as a meadow with a winding brook in the English countryside or an icy trail through a pass in the Alaskan mountains. It is difficult, almost certainly impossible, to describe a place so tellingly that the same scene will appear in the minds of different readers. Only a rough approximation is possible. Moreover, landscape often is used as a metaphor, in phrases such as "landscape of zeros." At a conference in Budapest I once heard the novelist E. L. Doctorow speak of the pages of his writing as a "landscape" which he could look at as a novel developed. Such literary uses of the term are not central to the meaning of landscape as it will be employed here, which is more strictly geographical and historical. Literary landscapes, like those in art, will be examined as an indicator of changing attitudes and values.

The anti-landscape has been a latent theme in the work of many cultural geographers and environmental historians. They have long been concerned with damage to landscape, for example because of land abuse, industrial pollution, or careless storage of hazardous waste. Much recent writing about environmental history has explored issues closely related to concept of anti-landscape, including Timothy McCain's *Mass Destruction: The Men and Giant Mines that Wired America and Scarred the Planet* (McCain, 2009) and David Zierler's *The Invention of Ecocide: Agent Orange, Vietnam, and the Scientists Who Changed the Way We Think about the Environment* (Zierler, 2011). Likewise, in April 2012 *Environmental History* devoted a special issue to the nuclear disaster at Fukushima. (Japan Forum, 2012)

In the worst cases, human beings have created places that cannot support life, and these are examples of the anti-landscape. Such places cannot sustain an ongoing human community within them. Whether temporary or long term, an anti-landscape suggests the post-human. Often these zones are the result of processes (such as desertification, erosion, deforestation, pollution or poisoning) that have occurred so slowly over generations that it can be difficult to grasp the long-term consequences of individual actions. In contrast, highly technological societies can create anti-landscapes in just a few years, for example at the reactors and plutonium processing plants in Hanford, Washington during and after World War II. (Findlay and Hevly, 2011, 14-73.) Some anti-landscapes are temporary, for example a town downwind of a chemical fire. Others are intractable and difficult to recuperate, such as the area around Picher, Oklahoma, where 170 million tons of lead and zinc mine tailings cover 7,000 acres. (Ed Keheley, et al., 2006) The concept of anti-landscape provides a point of unity for work that deals with sustainability, the limits to growth, dystopia, environmental degradation, "the tragedy of the commons," and the overuse of resources. It is specific enough to create a focused conversation but capacious enough to link with much current research, as exemplified by the essays gathered together in this volume.

The anti-landscape is not a new subject, but suggests a sharpening of interest in extreme forms of environmental degradation by giving them a collective name. George Perkins Marsh was thinking along similar lines in his seminal *Man and Nature*. He realized that settled societies were more destructive to the land than nomads. He saw that rich soils could be destroyed by deforestation.

> When the forest is gone, the great reservoir of moisture stored up in its vegetable mould is evaporated, and returns only in deluges of rain to wash away the parched dust into which the mould has been converted. The well-wooded and humid hills are turned to ridges of dry rock. . [the land becomes]. . an assemblage of bald mountains, of barren, turfless hills, and of swampy and malarious plains. There are parts of Asia Minor, of Northern Africa, of Greece and even of Alpine Europe, where the operation of causes set in motion by man has brought the face of the earth to a desolation almost as complete as that of the moon. (Lowenthal, ed., 1965, 42)

Since Marsh's time archeologists have documented how peoples have abandoned regions after they were stripped of trees, overgrazed, or too intensively irrigated. (See Ertsen in this volume.) Easter Island seems to be one clear case where excessive consumption of the local resources, notably through deforestation, degraded the habitat for human beings. (Ponting, 1992, 168-170) Thus Marsh's basic insight was confirmed by later work.

When a landscape has been degraded for generations, the damage becomes obvious to the unaided senses. People may respond to such places with dread, foreboding, and aversion. In *Landscapes of Fear* Yi-fu Tuan notes that "Fear is in the mind," but "has its origins in external circumstances that are truly threatening." (Tuan, 1979 6) This emotional response often arises directly from sensory impressions. In other cases, the fear is based on knowledge that our senses cannot detect. A dangerous chemical may be tasteless, a lethal gas may have no odor, and radioactivity is invisible. People have learned to be afraid of all three, but must rely on scientific measurements and experts to assess the danger. For example, in 2005 Hugh Kaufman, an expert on toxic waste, estimated that the chemicals carried into New Orleans by Hurricane Katrina would, as one newspaper reported, "make the city unsafe for full habitation for a decade." (Lean, 2005) Toxins had entered the floodwaters from the chemical plants and refineries along the Mississippi, commonly known as "Cancer Alley," which together produce 600 million pounds of toxic waste each year. As water evaporated from New Orleans, a residue of chemical sludge was left behind.

As the example of New Orleans during and after Hurricane Katrina suggests, one can think of anti-landscape in terms of three themes to which I will now turn: (1) the intentions of those who build a landscape and maintain it, (2) the dangers that arise at different historical moments, and (3) the specific geography of a location.

2. The Intentionality of Anti-landscapes

Intentionality offers a continuum of possibilities. At one extreme, some anti-landscapes came about without human intervention, and indeed pre-existed human beings. The surface of the moon cannot sustain human life. Volcanoes spew out lethal gases and spread lava over areas that for a long time afterwards cannot sustain life. In Cameroon a lethal cloud of CO_2 erupted out of a lake in 1986, killing more than 1,000 residents. This was not an isolated event. It can be expected to occur again, because carbon dioxide is continually being released into a number of African lakes. In Rwanda, for example, Lake Kivu is filling up with gas that seeps through the rocks beneath it. In early 2012 scientists estimated that it contained enormous quantities of CO_2, which bacteria have fed on to produce millions of cubic meters of methane.[1] The gases that are building up 250 meters or more below Lake Kivu periodically erupt to the surface, and can cause suffocation and explosions. (BBC, 2012) This naturally occurring, periodic disaster could transform the region inhabited by 2 million people around Lake Kivu into a deadly anti-landscape. To avoid this danger, the government has begun a test-program to extract the methane and pipe it to shore, where it can be burned to produce electricity. This is a good example of intervention to maintain a landscape so that it can continue to support human life.

The essays in this volume do not examine anti-landscapes that, like Africa's periodically exploding lakes, arise primarily through natural causes. Rather, they focus on unlivable places that came about either by design or as by-products of human activities. Some anti-landscapes are intended outcomes, notably in warfare. After the Third Punic War, it has been said the Romans plowed salt into the fields of Carthage to make certain their enemy would remain weak. It is not clear to what extent the Romans did this, but it would have undermined Carthage's agriculture, and with poor harvests the city would not rise again as it had after earlier defeats. Much more recently, the American military sprayed the defoliant Agent Orange over sections of Vietnam, to expose the enemy to view and to destroy their local ecology. Most anti-landscapes were created less intentionally, often accidentally. Some were only glimpsed as a remote possibility, like a factory or nuclear plant explosion, and usually backup systems and contingency plans were expected to prevent the release of radioactivity or to contain the effects of a chemical spill. In Bhopal, India such an accident occurred in 1984, when the insecticide methyl isocyanate rippled into the surrounding city from a Union Carbide factory. Almost 4,000 people died, more than in the 9/11 terrorist attacks. Moreover, in the following 25 years another 15,000 people have died from the aftereffects of the spill. In 2009 an estimated 10 to 30 people were still dying every month from exposure to tons of toxic waste still on the ground in Bhopal. Moreover, as *The New York Times* reported, "Dow Chemical, which has since acquired Union Carbine, refuses to accept any responsibility. The groundwater is contaminated; children of the survivors suffer from

1. An EU scientific team studied the gases dissolved in Lake Kivu. See http://mhalb.pagesperso-orange.fr/kivu/eg/eg_1h_janvier02.htm Viewed on May 14, 2012.

genetic abnormalities; and the victims" have no program that cares for them or that supplies even basic needs. (Mehta, 2009)

Intentionality is a matter of degree and it changes over time. The Bhopal disaster occurred because Union Carbide's safeguards were lax and buffer zones between the deadly chemicals and the urban population were woefully inadequate. Yet the accident strictly speaking was not intended but was due to negligence and poor planning. Years later, however, Dow Chemical knew it had acquired not just a company but its legacy of disaster, and it consciously chose not to take responsibility for the cleanup. Industrial pollution often begins as a thoughtless by-product, but over time becomes better understood, until eventually one can say that it has become intentional. In the nineteenth century factories and mills routinely dumped refuse into rivers and spewed dangerous substances into the air. In the twentieth century such pollution was gradually curtailed, as it was better understood. In this process, the unintended became intended, and exactly when this happened often became a legal question. For example, the Grace Company operated a mine just outside the town of Libby Montana. It did not willfully set out from the start to spread asbestos dust over the area. But after decades of mining, at some point company executives knew that they were filling the air with carcinogenic asbestos dust. Yet they continued business as usual. Certainly by the 1980s they understood that they were poisoning a whole population. (Sinclair and McCumber, 2004)

Similarly, the majority of anti-landscapes seem to begin as unintended outcomes, the by-products of often well-meaning projects. For example, the gradual drying up of the Aral Sea has turned the surrounding region into desert, shut down a fishing industry that once produced 50,000 tons per year, and undermined the life of what were once coastal towns now isolated in the desert. (Goudie, 1994, 202-206) This disaster occurred because the former Soviet Union diverted water from the Aral Sea's tributaries to irrigation projects. It took decades for the catastrophic results of this diversion to become obvious and nearly irreversible. By c. 2005 all but one of the fish species had died and the original ecological system was largely extinct. A now divided Aral Sea is partially recovering, as some tributary waters now reach it, particularly at the northern end, where a dam cuts off a section of the Aral Sea basin from the remainder, much of which seems fated to remain desert.

Unintended anti-landscapes are frequently caused not by a single decision but by a combination of circumstances and events. The modern factory, much less the modern city, contains so many overlapping and interacting systems, that unexpected accidents can occur due to a malfunction. In 1931 Stuart Chase described New York City as a potentially unlivable "Megalopolis" whose dust, noise, filth, and congestion were barely under control. "In brief, Megalopolis, for all its gaudy show, its towering architecture, its many refinements and cloistered comforts, is not physically fit for ordinary people to live in." He feared a technical breakdown of the physical city with disastrous effects on the inhabitants, the land values and the financial system. However, Chase also thought that if engineers and architects were given executive power to dynamite where necessary and rebuild, New York could be made "fit for the high gods to live in." (Chase, 1931, Chapter 3) As often happens, the same author both explores the dystopian possibilities of an anti-landscape and imagines a utopian future, where the best intentions are realized.

3. Temporality and Anti-landscape

A second aspect of the anti-landscape is that of time. It may exist for only a few hours or persist for generations. During a massive electrical blackout the possibility arises of a city becoming unsustainable. Even without a blackout some modern cities can become a death trap during a heat wave for people who lack air conditioning, as has happened, for example, in Chicago and Amsterdam. (Klinenberg, et al. 2003; Fischer, et al. 2004) But the viability of a city decreases in a blackout, to the extent that it cannot supply its inhabitants with light, food, and water. A blackout can be survived for a few days, but a few weeks, a month? Electrical utilities are adept at repair, and in most cases a blackout only provides a glimpse of the anti-landscape that lurks as a latent possibility should technological systems fail. The blackout typically lasts between a few hours and a few days, long enough for people to sense the danger without experiencing too much hardship.

The Aral Sea provides a different example, where the northern section has become a separate, recovering ecological system, while the southern portion remains a barren anti-landscape. There, the salty, pesticide saturated lakebed has dried into a pernicious powder that is picked up by winds that blow it into the towns along the old shoreline. Without the fishing industry and the life-giving waters, the area supports far less life than before. At least in theory, however, this part of the Aral Sea might recover if rainfall increases or irrigation decreases.

In contrast, the regions around Hanford and Chernobyl must be abandoned for a much longer time. They are fenced off and have been officially declared uninhabitable for generations. The zone around the 2011 nuclear accident in Japan has also been closed off, and its former residents have been relocated. Another example comes from irrigation, where some Californian fields, after years of evaporation, accumulated so much salt, selenium, arsenic, pesticides, and other toxins that they gradually became unable to support agriculture and poisoned the wildlife, and thus fit all too well the definition of anti-landscape. (Lopez, 1989, 29-36) Anti-landscapes thus can be ephemeral events that last only a few hours or days or they can last for generations.

4. Place and Anti-landscape

Anti-landscapes also vary greatly depending upon their physical location. Precisely the same toxic event can have quite different consequences, depending on where it occurs. An oil spill has longer-term ecological consequences if it occurs in a small arctic bay (where cold weather slows down the recovery process) than if an identical amount of oil spills in the middle of the warm and turbulent Gulf of Mexico. There is also a telling difference between sites with closed or permeable boundaries. Some sites have clearly defined boundaries, such as the nuclear waste storage facility that was carved out of the solid rock at Yucca Mountain in Nevada. It has hard and clear borders designed to contain nuclear waste within cathedral sized rooms. However, this site met objections from environmentalists and local residents and it never was taken into use. Instead, dozens of less secure "temporary" nuclear waste storage units remain in operation. Undesirable as this may be, they do not have boundaries as porous as the grounds of the petrochemical factories near New Orleans along the Mississippi River. Their chemical pollution can seep into groundwater and constantly moves into new locations. (Allen, 1999, 189-197) Likewise, nuclear radiation can be carried thousands of miles by the wind. The anti-landscape is a general concept, but every case is different, based on local circumstances. The Aral Sea has no outlet, and therefore rapidly became saltier once its tributaries were siphoned off for irrigation. In contrast, Lake Erie, a body of water smaller in size than the Aral Sea before it largely evaporated, receives water from three other lakes that is discharged through the Niagara River into Lake Ontario. Problems in any of the Great Lakes, such as pollution or invasive species, can move through the whole system. This, together with the general abundance of water creates quite distinct problems from those of the Aral Sea. Both have the potential to become an anti-landscape, but not the same kind, and for different reasons. Anti-landscapes may have hard or porous boundaries, which increase or decrease their interaction with the surrounding environment.

5. The Organization of this Volume

Each anti-landscape suggests a latent narrative with three parts: (1) there once was a healthier, sustainable place that supported life but was vulnerable; (2) whether by accident or design, that place was severely damaged and no longer could support human communities; (3) nevertheless, that place might be recuperated and again become a landscape in Jackson's sense of the term. This volume is organized in terms of these three narrative moments.

Part one's central concern is the sense of vulnerability in landscapes. While this sense of fragility and contingency can be found long before the twentieth century, as already discussed, physical vulnerability became a major cultural preoccupation during the early Cold War. First came fears of nuclear fallout (Weart, 1988) and an awakening to the dangers of chemical pollution (notably due to Rachel Carson's *Silent Spring*, 1962). By the 1970s, there was a more general awareness of what the Club of Rome called the "Limits to Growth," stirring up a debate that led many Europeans and Americans to adopt the view that economic growth could not be extended indefinitely. (Meadows, et al. 1972.) These concerns were gradually intensified by the discovery of global warming and the realization that human interference in the environment had become so endemic that it was possible to speak of the "end of nature." (McKibben, 2006)

The first article in this section, by James Fleming, examines how the ocean of air that we breath and which circulates constantly around the earth can be understood as both "skyscape" and "anti-skyscape." He notes that, "*Anti-skyscapes* can be infinitely more pernicious than even the most violent storm, which will surely pass. They were visible in the acidic skies of the Manchester factories, smog in London or Los Angeles, and dust storms over the prairies. More recently, however, they are largely invisible." Other dangers lie in the water supply, particularly in arid regions. Maurits Ertsen's essay delves into this subject, with a focus on the successes and failures of irrigation in the Indus River system in present-day Pakistan. Built during the British colonial administration, the system was modified in the last half century, with many tube wells added to make use of the rising water table level caused by irrigation. It is a complex story, not one of simple linear progression. Ertsen rightly emphasizes the necessity for a long historical perspective in assessing any agricultural system, and he reminds us that human creations such as an irrigation system can become unintended acts of slow-motion destruction.

Part one concludes with further thoughts about the anti-landscape, as Mark Luccarelli explores the possible meanings and implications of the term. He warns that it should not simply be considered as a binary opposition to landscape, but rather understood in political terms. To do this, Luccareli examines J. B. Jackson in historical and cultural context, locating his definition of landscape in an intellectual tradition. He then suggests ways to enrich that definition through the Lefebvre's work on space. Reading his work immediately after Ertsen's, makes good sense, because in quite different ways, each is writing about how landscapes are shaped by a combination of political, engineering, and economic decisions. Ertsen's work documents Luccarelli's assertion that "landscape is really largely a reflection of the social world, both in its boundary-making and use-dependent aspects."

Part Two turns to literary and photographic representations of the anti-landscape, beginning with Anne Flügge's discussion of Modernism in the 1920s and 1930s, when the image of the wasteland became a central trope in defining the western cultural condition. She begins with a review of T. S. Eliot's seminal 1922 poem, "The Wasteland," and shows how similar images and ideas suffused F. Scott Fitzgerald's *The Great Gatsby* and, in popular culture, the hard-boiled detective genre, exemplified by Raymond Chandler's *The Big Sleep.* Such despair seems almost genteel, however, when compared to the comprehensive devastation made possible by weapons of mass destruction. Nick Carraway does not imagine the entire New York area turned into a lifeless wasteland, but such visions became all too common in the early Cold War. Øyunn Hestetun focuses on a recent example of this genre, Cormac McCarthy's *The Road.* Taking place in an incinerated land where all forms of animal and bird life have died and only a few human beings wander the wasted world looking for useable fragments of its past, *The Road* presents an anti-landscape that does not sustain any form of life. Hestetun uses Lawrence Buell's concept of "toxic discourse" to explain the elements of McCarthy's novel, arguing that it is an eco-dystopia, a theodicy, an allegorical meditation on the grotesque, and a messianic quest. The novel is less a description of what a post-apocalyptic world would be like, than an allegorical exploration that necessarily simplifies the physical world to a dead backdrop. It is highly unlikely that any catastrophe would in fact destroy all insects, fish, birds, and animals and yet leave some human beings and the microorganisms in their digestive systems alive. Many species might become extinct, but unwanted insects survive inside nuclear reactors and many animals and plants seem to be thriving at the abandoned site of Chernobyl. But McCarthy's novel is not flawed because it ignores such likely survivals. His simplification of the biology of such a catastrophe sharpens the novel's focus on philosophical questions and human psychology.

Hannes Bergthaller helps us to read photographs of anti-landscapes, showing how the images of John Ganis in *Consuming the American Landscape* emerge out of and critique the nineteenth century landscape tradition, as represented by the work of Timothy O'Sullivan and William Henry Jackson and later reprised and made popular by Ansell Adams. Drawing on the work of the German philosopher Wolfgang Welsch, Bergthaller shows the usefulness of the concept of "anesthetic landscape" in understanding how each reconception of the landscape genre both opens our eyes and yet can blind us. This essay shows that the photographs in *Consuming the American Landscape* are more complex and interesting than they first appear, and that they are not just another expression of "the cult of wilderness" or another "lament for the raped virgin land." Rather, this imagery "marks the obsolescence of traditional landscape photography itself."

Werner Bignell next examines the possible aesthetic appeal of anti-landscape. While it might seem logical to assume that an unlivable space could only evoke revulsion, the response is often more complex. He argues, "mental images are projected onto the world not primarily for survival but as an expression of individual and collective engagement in the world. When this process fails, and the failure manifests itself in space, the result is an anti-landscape." The failure of such projections is therefore "part of biological and cultural adaptation to the world, a call to not confuse image and material reality and to learn from failure to create a better image; negation here is a corrective, not a predicament." From this perspective Bigell suggests the value of spaces like the American desert in the Southwest: "The desert is appealing because it transcends projection and the aim of projection, usefulness – it is literally a wasteland." In contrast, he considers American suburbia a failed adaptation to the environment and a landscape based on unsustainable consumption. Bignell argues "landscape and anti-landscape are dialectical twins" that define one another. Maunu Häyrynen complicates this dualism by looking at a phenomenon that is related to anti-landscapes, which he terms "lost landscapes." These are scenic places that have lost part or even all of their earlier cultural importance, whether due to political, social, or economic developments. The erosion of their cultural significance reminds us that sites may either fall into oblivion or have an afterlife in which they acquire new cultural meanings.

The third and final section turns to the question of the afterlife of anti-landscapes. To what extent can they be recovered and returned to sustainable use? This has happened more frequently than many people realize. Salt-infested irrigated fields have been recovered through various techniques. (Goudie, 1994, 145-150) The oil fields of Western Pennsylvania seemed to be devastated beyond recall in the 1860s and 1870s. The streams ran with oil, and the landscape was blackened beyond recognition. Today, the area is green, the streams have revived, and there are few signs of the environmental havoc that accompanied the oil rush. (Black, 2000, 64-67, 70-71, 76-77) Other sites have recovered from accidents and disasters, often with considerable government investment and scientific assistance. Many, perhaps most, anti-landscapes can be recuperated. The Cuyahoga River in Cleveland was an open sewer in the 1960s. In June of 1969 there was so much industrial sludge and oil-soaked debris floating on the surface that it caught on fire and burned until the fire department arrived to put it out. Indeed, the river had caught on fire many times before, and this was considered an ordinary occurrence in the 1930s. But by the 1960s such levels of pollution were no longer acceptable. The 1969 fire galvanized the public imagination and gave impetus to the creation of a national Environmental Protection Agency and to a new law, The Clean Water Act. Four decades and several billion dollars later, the river is clean. Sixty species of fish can be found in its waters, and fishermen can safely eat the bass and trout they catch there. Lake Erie as a whole is also becoming cleaner, its beaches safer, its fish more numerous. (Maag, 2009)

Other anti-landscapes were recovered in the United States, beginning in 1980, as billions of dollars were spent to clean up "Superfund Sites" such as Love Canal, New York. There, industrial waste had been buried in barrels that in the 1970s began to rust and leak into an area that had been developed into a Buffalo suburb. The pollution was so severe it forced evacuation of the residents and government purchase of their homes. A cleanup and containment program that lasted the better part of twenty years made the area safe, and its houses were refurbished and sold.

Anna Storm examines somewhat similar problems faced by two communities that grew up around industrial sites that have closed. One is a nuclear plant that was once the pride of Soviet industry, the other an open pit mine in Sweden. Each poses serious problems to its community, both culturally and environmentally. Nikolet Jensen provides a case study of how one derelict site, an abandoned railroad line on the West Side of New York City has been transformed into a linear, urban park. Not only has a deserted, crumbling piece of infrastructure been reborn as an attractive green zone, but the surrounding area has increased in value, and New Yorkers have a new way to experience their city from an elevated pathway.

In the afterword, Sarah Elkind reconsiders the concept of anti-landscape in light of these essays, offering as a final example the history of Stige Ø, where a small working-class community once lived. It was transformed twice. First, the local residents were evicted so it could be used as a landfill. Later, it was later capped and transformed into a recreation area. Such a site raises further questions about how to define anti-landscape.

Taken together, these essays stake out a terrain that is open for further research. They are concerned with the pollution or abuse of the earth, the sky, and the waters, and how these problems are interpreted and understood when they become threatening to human beings. The anti-landscape refers both to artistic and literary concerns and to specific places that are threatened or that no longer sustain life. The interdisciplinary work of the environmental humanities draws together scholars to study such sites, the terms of their destruction, the possibilities for their recovery, and their value in the aftermath of renewal.

Bibliography

Allen, Barbara. 1999. "Narrating the Toxic Landscape in 'Cancer Alley' Louisiana," in David E. Nye, ed. *Technologies of Landscape*. Amherst: University of Massachusetts Press.

BBC News Magazine, 2012. "Lake Kivu Gas," 13 Feb.

Black, Brian. 2000. *Petrolia*, Baltimore: Johns Hopkins University Press.

Carson, Rachel. 1962. *Silent Spring*. Boston: Houghton Mifflin.

Chase, Stuart.1931. *The Nemesis of American Business*. London: Macmillan.

DeLue, Rachael Ziady, and James Elkins, eds. 2008. *Landscape Theory*. London: Routledge.

Ed Keheley and Associates, Inc. 2006. "History of the Picher Mining Field," 40th Annual Meeting, The Geological Society of America, South Central Section, 6-7 March, 2020 South 640 Road, Quapaw, Oklahoma 74363.

Findlay, John M and Bruce Hevly. 2011. *Atomic Frontier Days; Hanford and the American West*. University of Washington Press.

Fischer, P. H., B. Brunekreef, and E. Lebret. 2004. "Air pollution related deaths during the 2003 heat wave in the Netherlands." *Atmospheric Environment* 38.

Goudie, Andrew. 1994. *The Human Impact on the Natural Environment*. Fourth Edition. Cambridge, MA: MIT Press.

Jackson, J. B. 1994. *A Sense of Place, A Sense of Time*. New Haven: Yale University Press.

Jackson, J. B. 1984. *Discovering the Vernacular Landscape*. New Haven: Yale University Press.

Japan Forum. 2000. *Environmental History* 17:2 (April), 217-300.

Klinenberg, Eric. 2003. *Heat Wave: A Social Autopsy of Disaster in Chicago*. Chicago: University of Chicago Press.

Lean, Geoffrey. 2005. "Cover-Up: Toxic Waters 'will make New Orleans unsafe for a decade.'" *The Independent*, 11 Sept.

Lopez, Barry. 1989. *Crossing Open Ground*. New York: Vintage.

Lowenthal, David ed. 1965. George Perkins Marsh, *Man and Nature*. University of Washington Press.

Maag, Christopher. 2009. "From the Ashes of '69, a River Reborn," *New York Times*, June 21.

McCain, Timothy. 2009. *Mass Destruction: The Men and Giant Mines that Wired America and Scarred the Planet*. New Brunswck: Rutgers University Press.

McKibben, Bill. 2006. *The End of Nature*. New York: Random House.

Meadows, Donella H., Dennis L. Meadows, Jorgen Randers, and William W. Behrens III. 1992. *The Limits of Growth*. London: Pan Books.

Mehta, Suketu. 2009. "A Cloud Still Hangs Over Bhopal," Op-Ed column, *The New York Times*, Dec. 3.

Nye, David E. 2007. "Are Blackouts Landscapes?" *American Studies in Scandinavia* 39:2, 72-84.

Nye, David E. 2010. *When the Lights Went Out.* Cambridge, MA: MIT Press.

Ponting, Clive. 1992. *A Green History of the World*. Hammersmith: Penguin Books.

Schneider, Andrew and David McCumber. 2004. *An Air that Kills: How the Asbestos Poisoning of Libby, Montana, Uncovered a National Scandal.* New York: G. P. Putnam's.

Tuan, Yi-Fu. 1979. *Landscapes of Fear.* New York: Pantheon Books.

Weart, Spencer. 1988. *Nuclear Fear.* Cambridge, MA: Harvard University Press.

Zierler, David. 2011. *The Invention of Ecocide: Agent Orange, Vietnam, and the Scientists Who Changed the Way We Think about the Environment*. Athens, GA: University of Georgia Press.

I

The Threatened Landscape

2

Skyscapes and Anti-skyscapes: Making the Invisible Visible

James Rodger Fleming

A skyscape is a view of the sky or an artist's depiction of part of it. It is often strikingly beautiful, evanescent, and colorful, but it may also be an anti-skyscape – gloomy or menacing. Undoubtedly the genre has received less attention from scholars than its landed cousin. This essay examines the long history of skyscapes and anti-skyscapes and how recent collaborations between artists and environmentalists are making even invisible aerial threats visible.

Keywords: carbon emissions; climate; environmental art; landscape; painting; pollution; representation; sky-scape

1. Introduction

Approximately half of every landscape painting, from the horizon up, is a skyscape, sometimes celebrated, often largely ignored, but unquestionably extremely difficult to render in most any media, given its transience and transcendence. Many are realistic, as in the work of Albert Bierstadt, John Constable, and George Innes; others, by George Catlin, Claude Monet and Alfred Stieglitz, are smoky and moody; recent works by Andrea Polli and HeHe are abstractions. The vista of a natural skyscape, whether serene or ominous, is undoubtedly evanescent, rendered visible by reflected and refracted light. Blue sky and white clouds are only two of myriad colors that include the black of storm clouds and the near infinite luminous pastels of sunrise and sunset, rainbow, aurora, and other meteors.[1]

1. I want to thank Nicole Sintetos, STS Research Assistant and Lauren Lessing, Mirkin Curator of Education, Colby Museum of Art for their assistance with this research.

Anti-skyscapes can be infinitely more pernicious than even the most violent storm, which will surely pass. They were visible in the acidic skies of the Manchester factories, smog in London or LA, and dust storms over the prairies. More recently, however, they are largely invisible. They are the polluted spaces of our planet, including now the planet itself, often rendered visible through idealized false color scientific imagery. Anti-skyscapes undergird narratives of environmental degradation, anxiety, despair, disease, dystopia, collapse, resilience, sustainability, collective action, inaction, heroic action, and even faith. Miasmas. Acid rain. The ozone hole. Carbon dioxide climate warming. No one has seen them, yet, in a way, everyone has. Today, highly politicized images of the atmosphere link scientific, environmental, technological, and aesthetic histories as they codify and symbolize fear, angst, urgency, control, and lack of control.[2]

This paper mobilizes cultural studies, science studies, and art history to examine human-modified, but ubiquitous aerial spaces that are both ominous and threatening, yet truly serve, across boundaries, as the "background for our collective existence." (Jackson, 1984) The genre of skyscapes complements the terrestrial conversation and highlights emerging interests in environmental art, toxic airs, and climate engineering. They are interdisciplinary, international, and intergenerational vehicles.[3] Skyscapes are often overlooked. Anti-skyscapes are overlooked at our peril. Look up. And look out!

2. Skyscapes and Landscapes

"Sky" is the apparent arch or vault of heaven, whether covered with clouds or clear and blue; it may be the climate or clime of a particular region,

---

2. This was the theme of the conference "Pictur(e)ing Climate – Visualizations, Imaginations, Documentations," Institute for Arts and Media, University of Potsdam and Potsdam-Institut für Klimafolgenforschung (PIK), Potsdam, Germany, January 2012.
3. For example Constable's work is interdisciplinary, linking art and science, Poli and HeHe convey global messages, and, as is the case for many artists, Steiglitz's photographs are interpreted differently by different generations. A new article by Polli places her work in conversation with historians of chemistry and environmental historians: "Who Owns the Air? The Buying and Selling of Greenhouse Gases," in *Toxic Airs: Chemical and Environmental Histories of the Atmosphere*, ed. James Rodger Fleming and Ann Johnson (Pittsburgh: University of Pittsburgh Press, in press).

nowadays usually designated more globally than locally. The appearance of the sky is variously sunny, starry, hazy, overcast, azure, copper, even milky white. According to Raffaele Milani, sky is "a concept that incorporates meteorology, astronomy, and astrology, and we find within it theological theories pertaining to the origins of the cosmos. The creation myths describe the marriage of earth and sky. In the Bible the sky is the throne of God and is represented as divided into levels and vaults, which are the seats of angels of different orders. In ancient Chinese symbolism, instead, the sky represents the energy moved by the destiny that directs all things terrestrial; it is a cosmological image in which the sky does not appear as a symbol of the afterlife." (Milani, 2009, 141) The sky has been read, historically, as a panorama of visible portents. It also carries the breath of life and the weather each day, whether gently cooling or wildly devastating winds, nurturing or flooding rains, or life-giving warmth or enervating heat.

Medieval artists made God's wrath visible (Illustration 1). "Then the LORD rained on Sodom and Gomorrah sulfur and fire from the LORD out of heaven" (Genesis 19:24). The *ménage-à-trois* (literally) in the foreground depicts Lot with his daughters, who are making him drunk

Illustration 1. Master of the Lille Sermon, The Burning of Sodom, Flemish, ca. 1550-1575, oil on panel. Bowdoin College Museum of Art, Brunswick, Maine. http://www.bowdoin.edu/art-museum/exhibitions/2009/art-history-100.shtml

so he will sleep with them, which they do (Genesis 19:30-38). We can assume that the Biblical authors and the medieval artist did not have air pollution in mind. Nevertheless it is an anti-skyscape in which the invisible is being rendered visible and the numinous threatening.

The English Romantic painter John Constable is widely known for his bucolic country landscapes, with more or less naturalistic clouds painted under the scientific influence of the contemporary London scientist Luke Howard, who also influenced Goethe. This seascape with rain clouds (Illustration 2) is particularly dynamic, since the sky occupies 80 percent of the frame, and the clouds and rain are particularly powerful and eventful, depicted by bold and aggressive brush strokes reaching from the sky to the choppy, dark sea.

Illustration 2. John Constable, English (1776-1837). Seascape Study with Rain Clouds. 1824-5, oil on paper mounted on canvas, 220 x 310 mm. University of California, San Diego. http://en.wikipedia.org/wiki/File:Constable_-_Seascape_Study_with_Rain_Cloud.jpg

Humans entered the skyscape in 1783 based on the pioneering work of *les Frères* Montgolfier (Joseph-Michel and Jacques-Étienne) in scientific ballooning, but the balloons were soon applied to military purposes. In

The Balloon, or The Rising of the Montgolfiere (circa 1816), by Spanish painter Francisco Jose de Goya y Lucientes, a Montgolfier-style a balloon is soaring over the battlefield, higher than the mountain and higher than the clouds. The signal flags indicate the aeronauts are actively involved in the combat below, where death is crammed into the hollows of the landscape. The general (Napoleon?) on the white horse and the observers on the hillside have no such privileged perspectives. It is an anti-landscape depicting the new technology of aerial surveillance that made the devastation of war more complete.*

Photography brought new scientific, technical, and artistic opportunities and challenges in rendering the sky. Gustave Le Gray (1820-1882), a noted French photographer known for his images of the Mediterranean

Illustration 3. Gustave Le Gray, French (1820-1882). An Effect of the Sun, Normandy. 1856, albumen print, 32cm x 41.8cm. Cleveland Museum of Art, Cleveland, Ohio, via the Amica Library. http://amica.davidrumsey.com/luna/servlet/view/all/who/Le+Gray,+Gustave/where/European?cic=AMICO~1~1&os=0&pgs=50&sort=INITIALSORT_CRN%2COCS%2CAMICOID

* Not depicted, but held by the Musée des beaux-arts, Agen, Lot-et-Garrone, France.

Sea, developed a new methodology for toning prints in order to capture the evanescent lighting of both the sea and the sky. A double exposure of an albumen suspension captured two elements separately, with the negatives combined to produce a final image, as in the 1856 print shown here. The skyscape is first (literally) separated from the landscape, then visually combined. How much time has elapsed between the two exposures?

Scott Peterman's C-print on paper, Sao Paulo (2003) is a contemporary example of a cityscape/ skyscape double exposure that illustrates how artists are interpreting urban growth. The sky is manipulated in this photograph until it almost disappears. As your eye is drawn to details of buildings sprouting up near the horizon, the sky reappears as a heavy mood brooding over the city, lacking in all detail, but illuminating nonetheless the surging white towers. It is the absence of the sky that makes its presence felt and makes it so pronounced here (Illustration 4).

Illustration 4. Scott Peterman, American, (b. 1968). Sao Paulo. 2003, C-print on paper, 20 in. x 24 in. Colby Museum of Art, Colby College, Waterville, Maine. http://www.colby.edu/academics_cs/museum/search/Obj5351?sid=2207&x=17999

The technology of aerial flight has entered the skyscape in this 1910 Stieglitz print, lofting the aeronauts in their heavier-than-air craft far above the cares of this world. Are they really invincible? Or has aviation made us all the more vincible? Four years before the onset of the Great War, the militaries of the world had already become the leading contractors for aeroplanes as war machines, changing the meaning of the sky, and making it ominous for reasons other than storms.

Illustration 5. Alfred Stieglitz, American (1864-1946). Aeroplane. 1910. Photogravure on paper, 5 in. x 6 ¾ in. Accession Number: 1974.102. Colby Museum of Art, Colby College, Waterville, Maine. Gift of Norma B. Marin. Image source Museum of Modern Art. Transferred from the Museum Library. © 2011 Estate of Alfred Stieglitz / Artists Rights Society (ARS), New York. http://www.moma.org/collection/browse_results.php?object_id=51519

Art critics discussing landscapes often don't really "see" skyscapes. At a recent gallery talk at the Colby College Museum of Art, a professor of art discussed Albert Bierstadt's *View of Chimney Rock* (1860) for upwards of an hour, with nary a mention of the upper seventy percent of the image. This painting (Illustration 6), a promotional and commercial piece in its day, depicts the sun setting over the landmark Chimney Rock on the Oregon trail and a scene of domestic tranquility in life of the Ohalilah Sioux, indicating, according to the professor, the sun setting on their way of life too. But for weary travelers from the East reaching Chimney Rock, Nebraska, the meaning was different. They were on the trail of Manifest Destiny, and the rock marked officially the beginning of the American West, where their destiny lay. In this case the vast expanse of the sky signified to Bierstadt the seemingly limitless potential for American expansion as it invited ever more settlers into the West.

Illustration 6. Albert Bierstadt, American (1830-1902). *View of Chimney Rock, Ohalilah Sioux Village in the Foreground.* 1860, oil on board, 13¼ in. x 19 3/8 in. Colby Museum of Art, Colby College, Waterville, Maine. http://www.colby.edu/academics_cs/museum/search/Obj1743?sid=2207&x=18083

In Frederick Church's hardly discrete painting of the Unionist energies of the 1860s (Illustration 7), stars in a field of deep blue and the fiery colors of the sunset (or is it the sunrise?) echo the national angst and determination. A dead snag becomes a flagpole and portents in the sky become symbols of national unity across a devastated landscape. The smoke and fire of distant battle might well contribute color to the scene.

Illustration 7. Frederick Edwin Church, American (1826- 1900). Our Banner in the Sky. 1861, oil on paper, 19.21 cm x 28.89 cm. New York State, Office of Parks, recreation and Historic Preservation, Olana State Historic Site. http://www.the-athenaeum.org/art/full.php?ID=26490

3. The Threatening Skies: Visible and Invisible

During the smoky industrial revolution (Lewis Mumford's paleotechnic era), the atmosphere was visibly polluted. "The global-scale transformation of the environment by industrialization was… nowhere more evident than in the atmosphere."[4] Yet for an industrialist it was not threatening, the smoke of industry ca. 1900 represented prosperity. This photogravure (Illustration 8) by Alfred Stieglitz, "The Hand of Man," appears to our eyes as a highly polluted example of human degradation of the environment, however, Stieglitz and his circle of artists saw American industrialization as a measure of its strength.

4. Will Steffen, Paul J Crutzen, and John McNeil, "The Anthropocene: Are Humans Now Overwhelming the Great Forces of Nature?" *Ambio* 36, no. 8 (2007): 614-621, on 616.

Illustration 8. Alfred Stieglitz, American (1864-1946). Hand of Man (Long Island City). 1902, Photogravure, 8 3/4 in. x 6 5/8 in. Colby Museum of Art, Colby College, Waterville, Maine. http://www.colby.edu/academics_cs/museum/search/Obj2643?sid=2207&x=18042

But what about the threatening sky we can't see? "The wind and weather, ever on the move, afflicted the body not only by penetrating it directly, but also by corrupting the local environment in which a body dwelled."[5] This quote, by Brenda Gardenour, describes bodily exposures in antiquity and the Middle Ages, but it also applies to later eras. In the eighteenth and nineteenth centuries miasmas propagated from the stench of cities; night airs were to be feared; and, in an age of exploration and colonization, the airs of a foreign country were never as good as those of the homeland. But new miasmas afflict us in the twentieth and twenty-first centuries. The factory brought visible smoke, unhealthy particulates, and acid rain; the auto too transformed the blue skies over cities into brown clouds of photochemical smog, while emitting invisible carbon monoxide locally

5. Brenda Gardenour, "*Aer Corruptus et Noxio Humore*: Corrupt Air, Poisonous Places, and Toxic Breath in Late-Medieval Medicine and Theology," in *Toxic Airs*, ed. Fleming and Johnson (in press).

and invisible carbon dioxide globally. Horrible poison gas clouds rolled over battlefields and choked both foe and ally, while clouds of tear gas still accompany protests in city streets.

The atomic bombing of Hiroshima generated a new phenomenon – black rain, the remains of an incinerated city falling back to earth, but it also released the nuclear alchemy of invisible fallout on the world. Bomb blasts over the deserts and oceans soon gave way to bomb blasts in space designed to disrupt Earth's electrical and magnetic environment.[6] This was not the only alchemy, however, for the subtle but persistent cloud seeding agent silver iodide, designed to "trick" a cloud into precipitating, soon became a hoped-for means for the military to control the sky.[7] As cool breezes of air-conditioning filled our homes and workplaces, seemingly inert and invisible CFCs began their devastating ascent into the stratosphere. Now, because of stratospheric ozone depletion, children are taught to fear exposure to the clear blue sky. How many artists of earlier eras have depicted this emotional response?

4. Contemporary Skyscapes

There is a synergy between environmental history and contemporary art. The work of eco-feminist artist, Janet Culbertson, transforms a metaphor of "sky as commodity" into an ecological message. Known for her poignant, futuristic political paintings from the 1970s to the present, Culbertson's billboard series is an ironic self-reflexive representation of our time, in which the ideal landscape is ominously projected off of a billboard. Her critique of the dominant anthropocentric ideology behind nature and consumption allows the viewer to question the security and stability of the skyscape itself.

In her work *Vanishing Gold* (Illustration 9), the brilliant skyscape becomes contained and reflected by the equally radiant water spilling off the billboard's side. Our gaze is drawn to the orange and yellow sky

6. James Rodger Fleming, "Iowa Enters the Space Age: James Van Allen, Earth's Radiation Belts, and Experiments to Disrupt Them," *Annals of Iowa* 70 (Fall 2011), 301-24.
7. James Rodger Fleming, *Fixing the Sky: The Checkered History of Weather and Climate Control* (New York: Columbia University Press, 2010).

and water of the idealized, commodified scenery juxtaposed against the distant pinkish sky. Culbertson problematizes our conception of "natural beauty" since the colored faux skyscapes are likely due to high levels of particulate matter or some other foreign matter in the air or chemicals in the water.

Illustration 9. Janet Culbertson, American (b. 1932). Vanishing Gold. 1998, oil on canvas, 22 x 30 in. Collection of the Artist. http://www.janetculbertson.net/bill2.html

As was the case at the dawn of photography, art continues to evolve with technology, and new modes of depicting the sky become possible. Contemporary artists Professor Andrea Polli of the University of New Mexico and the French duo HeHe are expanding the boundary between science, technology, and environmental art. Polli has often collaborated with atmospheric scientists to depict the politics and reality of climate change and pollution through art.[8] "Particle Falls," completed in collaboration with Chuck Varga, is an urban, environmentally-motivated piece that renders "the invisible visible" (Illustration 10). Placed on a city corner

8. Andrea Polli, http://www.andreapolli.com/

in San Jose, an interactive sensor alters the waterfall depending on the amount of particulate matter in the air and can sense even a puff from a cigarette or fumes from a passing bus. Although we all know about air pollution, how real is it to us? By its invisible nature, do we often remain in denial about air quality until the smog that hits us is unbearable or a cough is insuppressible?

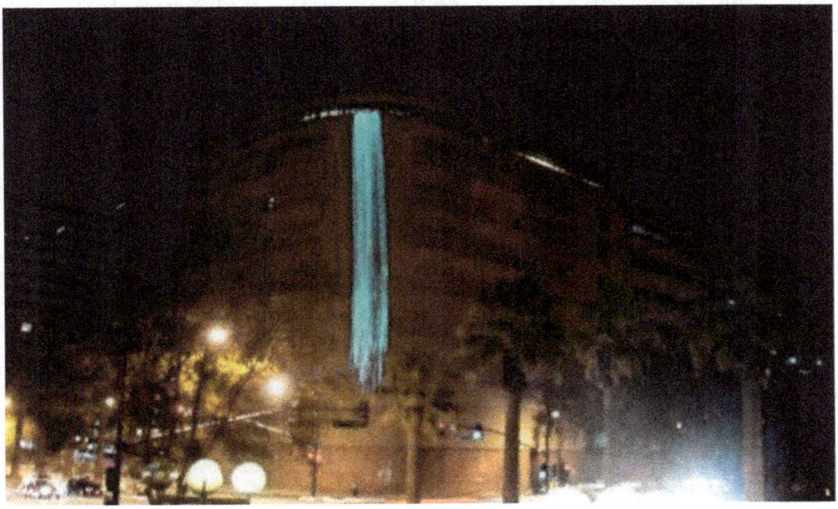

Illustration 10. Andrea Polli, American. Particle Falls (2010), installed in San Jose, California. Video is at http://vimeo.com/16336508

HeHe, a collaborative French art duo, Helen Evans and Heiki Hansen, have been using tactics similar to Polli in order to extend the dialogue on eco-consciousness to a larger scale. HeHe's work, "Nuage Vert", or "Green Cloud" (Illustration 11) uses lasers to create a green outline of the pollution emitted from Helsinki's Salmasaari power plant. Although projected into the sky, the critique of "carbon emissions" is decentered from being solely an atmospheric malady, but a complex and interconnected system mixing earth, wind, energy, and sky. HeHe notes:

> The physical dimension of the Salmisaari site, inhabits a special position in Helsinki; physically, visually and metaphorically. The vertical vectors of the architecture connect the underworld with the sky: the underground coal storage tunnels descend to 126m

below sea level (the deepest point in Helsinki) whilst its chimney reaches 155m into the sky with its cloud disappearing into the lower atmosphere. [9]

The skyscape then becomes not a symbol of industry, progress, or even nationalism, but the realization and template for the interconnectedness of both terrestrial and atmospheric pollution. More than a naturalistic horizon, the mobility of air itself decenters it from a simple point source, and in this way, political projects can be seen to apply to individuals and cities, not just to the policy makers.

HeHe's work is ironic, in the sense that it attempts to render aesthetic, using laser light, a human-made cloud, yet is also incredibly complex for blurring the line between art and activism, technology and aesthetics, and the public and private spheres. The project was collaborative in the sense that HeHe worked with owners and managers of the power plant to gain data on energy consumption, with scientists to configure the actual lasers, and with the Helsinki public through a poster and sticker campaign for an "unplug" event in which citizens reduced their power intake. As the levels of energy use were documented to decrease, HeHe increased the size of the laser cloud to show their progress. The sky then becomes the ultimate, unifying template to spread a public message, for as Helen Evans notes, "It shifts the discourse about climate change and carbon emissions from abstract immaterial models based on the individual, to the tangible reality of urban life."[10] The collaborative method also becomes a model for policy change, as artists create an equally aesthetic and social message, concentrating on the need for a community effort (the public to reduce consumption, the policy makers to push for greener bills, scientists to create greener technology) in order to enact social change and promote environmental protection.

9. "Gallery: Nuage Vert, Green Cloud…," http://inhabitat.com/green-cloud-hehe-helsinki-environmental-art/green-cloud-2/
10. Helen Evans, "Nuage Vert" *Cluster Magazine*, Issue 7. Torino, May 2008, in http://hehe.org.free.fr/hehe/texte/nv/index.html

Illustration 11 . HeHe, French. Nuage Vert, February 2008. http://hehe.org.free.fr/hehe/NV09/index.html, *links and video at* http://hehe.org2.free.fr/?language=fr

While always of interest, or at least worthy of a glance once in a while, clear, cloudy, and polluted skies have taken on new aesthetic life and meanings, supported in part by new collaborations between artists and environmentalists, including environmental historians. Still, much is yet to be done when it comes to depicting and interpreting skyscapes and anti-skyscapes. As artists are wont to remind us, it is the lighting of the sky that illuminates the rest of the landscape.

Bibliography

Bermingham, Ann. 1986. *Landscape and Ideology: The English Rustic Tradition, 1740-1850*. Berkeley: University of California Press.

Damisch, Hubert.2002. *A Theory of /Cloud/: Toward a History of Painting*. Transl. Janet Lloyd. Stanford: Stanford University Press.

Evans, Helen. 2008. "Nuage Vert" *Cluster Magazine*, Issue 7. Torino, May, http://hehe.org.free.fr/hehe/texte/nv/index.html

Fleming, James Rodger. 2010. *Fixing the Sky: The Checkered History of Weather and Climate Control* (New York: Columbia University Press.

Fleming, James Rodger. 2011. "Iowa Enters the Space Age: James Van Allen, Earth's Radiation Belts, and Experiments to Disrupt Them," *Annals of Iowa* 70, 301-24.

Fleming, James Rodger and Ann Johnson, ed. *Toxic Airs: Chemical and Environmental Histories of the Atmosphere*. Pittsburgh: University of Pittsburgh Press, in press.

Jackson, J. B. 1984. *Discovering the Vernacular Landscape*. New Haven: Yale University Press.

Milani, Raffaele. 2009. *Art of the Landscape*. Montreal: McGill-Queens University Press.

Miller, Angela L. 1003. *The Empire of the Eye: Landscape Representation and American Cultural Politics, 1825-1875*. Ithaca: Cornell University Press.

O'Toole, Judith H. *Different Views in Hudson River School Painting*. New York: Columbia University Press in association with Westmoreland Museum of American Art, 2005.

Reed, Arden. 1983. *Romantic Weather: The Climate of Coleridge and Baudelaire*. Hanover: Published for Brown University Press by University Press of New England.

Schlossman, Jenni L. 1992-1993. "Janet Culbertson: Political Landscapes," *Women's Art Journal* 13: 27-31.

Steffen, Will, Paul J Crutzen, and John McNeil. 2007. "The Anthropocene: Are Humans Now Overwhelming the Great Forces of Nature?" *Ambio* 36, no. 8: 614-621.

Thornes, John E. 1999. *John Constable's Skies: A Fusion of Art and Science*. Birmingham, UK: Birmingham University Press.

3

Step after Step the Ladder is Ascended Human agency in Irrigated (anti) Landscapes

Maurits W. Ertsen

Irrigation is an obvious blessing that often disguises long-term problems. It provided food for ancient civilizations by improving growing conditions for crops, but could cause crop failure by increasing soil salinity or vulnerability to climate change. In the modern Indus Basin in Pakistan irrigation may be heading for disaster as groundwater levels drop dramatically. In the early twentieth century, Indus Valley groundwater levels rose because of large-scale irrigation development. About fifty years ago, tube wells were installed to lower groundwater levels. These same tube wells are now blamed for depleting the groundwater. Perhaps the anti-landscape being created today is the old, dry landscape of the eighteenth century. When taking a closer look at the Indus Basin, a period of some two centuries shows the importance of human agency in anti-landscape creation. In the Indus, one encounters human construction of irrigated landscapes, their potential turn into anti-landscapes, and the human responses creating other, better irrigated landscapes. Studying the Indus yields new, interesting questions in the debate on the collapse of ancient civilizations – and potentially our own.

Keywords: agriculture; Climate Change; Indus River; irrigation; Pakistan; reclamation; salinity; tube wells; water

1. Introduction

Irrigation can bring wealth and stability to communities and nations; in fact it was irrigation that enabled most of the ancient urban societies to develop. However, in the long run, irrigation can also harm landscapes and food production. The Indus River watershed contains one of the largest irrigated areas in the world. With its reservoirs, barrages and main canals, the Indus Basin Irrigation System in Pakistan serves an area of 16 million hectares with some 172 billion m^3 of river water per year (Aslam and Prathapar 2006: see Illustration 1). This irrigation system may be heading for disaster, as groundwater levels drop dramatically. Although the current crisis is real enough – Pakistani farmers are already experiencing increased problems sustaining agricultural production – the irony of human-induced changes in groundwater levels is hard to miss. During the first half of the twentieth century, large-scale irrigation caused Indus groundwater levels to rise, threatening agricultural production. Around 1960, tube wells were installed to lower groundwater levels; these same tube wells are now blamed for depleting the groundwater and thus threatening agriculture once again.

This essay examines the creation of irrigated landscapes in the Indus Basin and the region's possible future as a non-irrigated, anti-landscape. It presents irrigation development as a series of human actions that modified the natural environment, with production benefits in the short run but potentially negative consequences for both human beings and the land when considered over decades or centuries. Examples from the Indus Valley suggest that a narrow environmental threshold separates the stable, irrigated landscape from unstable anti-landscapes. Once a society crosses that threshold, irrigation systems become more vulnerable, as relatively small additional changes can cause disaster. Because irrigated anti-landscapes result from complex human-environment interactions, however, analyzing the rate at which societies cross these thresholds, and even determining the importance of human agency in this kind of environmental transformation is extremely difficult.

Contemporary society is not the first to face irrigation-related problems. The archaeological record is filled with irrigation-related disasters, many of them related to climatic change. However, these disasters usually unfolded over centuries, and this has limited scholars' ability to incorporate human agency into their analyses. Improved understanding of human construction of irrigated landscapes, their potential transformation into anti-landscapes, and the role of human agency yields new insight into the collapse of ancient civilizations – with clear implications for our times. The transformations underway in the Indus Valley appear, on the surface at least, comparable to the patterns of societal collapse which Jared Diamond attributed to the over exploitation of natural resources in *Collapse: How Societies Choose to Fail or Succeed* (2005). Archeological literature discusses numerous cases where societies that depended upon irrigation ultimately "collapsed." These include ancient Mesopotamia that would have collapsed because of salinization due to over-irrigation, and the Hohokam in the American Southwest that would have collapsed because their dependence on irrigation prevented this society from responding to climate change (Howard 1993). Even so, critics have challenged Diamond's whole notion of societal collapse, first by challenging Diamond's historical accuracy (McAnany and Yoffee 2010), and then by arguing that these societies have not so much *collapsed* as *changed* (Schwartz and Nichols 2006).

To these critics, I would also add that any discussion of societal collapse should incorporate the time or the rate of change. Many of the civilizations that Diamond cites collapsed or disappeared over the course of centuries or millennia – periods of time during which societies should be allowed to change without being labeled as unsustainable. The perhaps climactically induced 4.2 K and 8.2 K events are a case in point (Akkermans et al 2010; Cheng at al 2009; Hijma and Cohen 2006; Weiss et al 1993). Are these to be understood as taking place over such time frames? New evidence from carbon dating (Akkermans et al 2010), however, seems to suggest that some of the most dramatic of these changes unfolded over the course of mere decades. As climatic changes typically do not emerge in timescales as short as decades, this new data suggests that something other than climate may have been at work in these cases. Clearly, dependence upon irrigation has rendered some societies more environmentally vulnerable in the long run, but just as importantly, the time scales of these social transformations matter greatly both in the analysis of collapse and environmental instability on one hand and in understanding the role of irrigation practice and climate change in the creation of anti-landscapes on the other.

2. Irrigation in the Indus Basin in Pakistan

In the Indus area, Pakistani farmers irrigate with both surface water from canals and groundwater from tube wells, creating a mixed pattern of public canal water and private tube well water distribution and control (see Illustrations 2 and 3 for impressions). Of the total water available at the farm gate in 2007-08, about 35% was contributed by groundwater (Agricultural Statistics of Pakistan, 2008-09). The Punjab is not only the main food-producing province in Pakistan; it also has the most tube wells and the largest population in the country. In many areas the groundwater contribution to irrigation exceeds 50%; as a whole, farms in the Indus Valley meet 40% of their irrigation needs with groundwater. Moreover, the use of groundwater from tube wells for irrigating crops has grown dramatically in the last 50 years. In the 2008-09 season, the number of registered tube wells, both public and private, was just over 870,000 (Agricultural Statistics of Pakistan). In the early 1950s the number could not even be counted in the thousands. The effects are as could be expected: groundwater levels have dropped significantly in recent years.

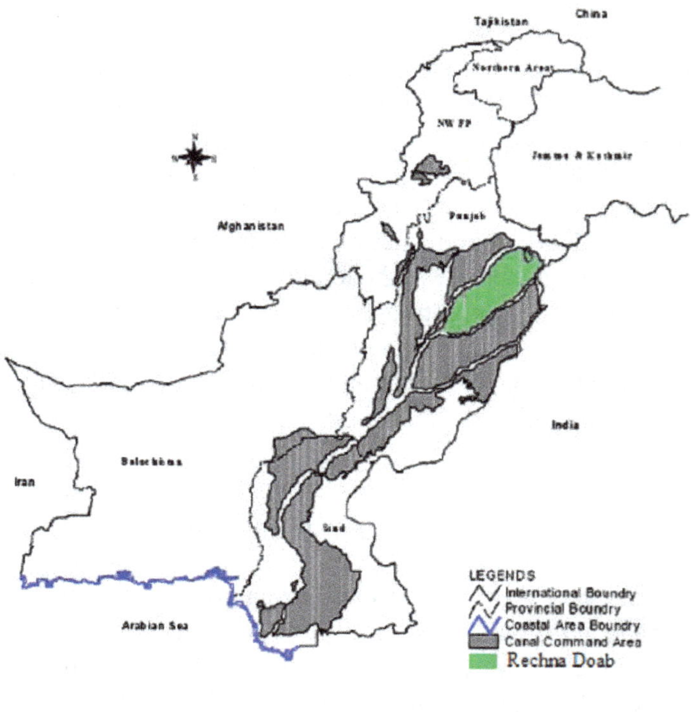

Illustration 1. The Indus Basin irrigation area (Ullah et al. 2001)

Illustration 2. Typical field irrigation infrastructure in the Indus Basin

Illustration 3. Poorly maintained delivery canal in the Indus Basin.

What does the growth of groundwater-based irrigation in the last 50 years suggest? Are the Indus farmers and managers heading towards collapse, a term very popular in the popular and scientific discourse? (Abbott 2003; Costanza et al 2007; Drysdale et al 2006; Fisher et al 2009; Staubwasser and Weiss 2006; Stone 2006; Weiss et al 1993). An archeological view of the Indus irrigation system reveals material evidence of irrigation, including canals of different sizes serving larger and smaller areas, tube wells and settlements. These structures must be examined in light of both the dates of their construction and the climactic data for the period from 1900 to 2000. Apart from a large number of tube wells, material remains of irrigation in the Indus Valley typically consist of tube wells that deliver water to an open canal system that carries water directly to fields. Furthermore, several larger canals are connected to these field canals, bringing water from rivers and larger supply canals. The field canal system shows signs of much better maintenance than the larger canals upstream in the system. The poor maintenance of the larger canals suggests that the water from rivers and the river-fed canal system is less important to Indus Valley farms than field canals and tube well water. Why the water delivery through the surface canal system should have declined in local importance is not clear, but it suggests that water availability through the river-fed irrigation system has decreased.

The conditions that inspired the Indus Valley irrigators' quest for ever more water from tube wells to supplement or replace surface supplies may have been the result of climatic changes in the region. The IPCC's shows that global temperatures have increased in the last 50 years or so, and will continue to rise in the future. Pakistan has had temperature increases similar to the global mean. Furthermore, it appears likely that summer hot spells in East Asia have – and will – become, more intense and frequent, and that there will be fewer very cold days (Solomon 2007). In other words, irrigation would have become more important due to these climatic changes, even though the IPCC also suggests that precipitation may increase. Computer models suggest that any increased precipitation in this region will occur not in the form of gentle rain but as cloudbursts and other intense precipitation events.

When we take these two issues together, we see that between 1950 and 2000, the number of tube wells increased as global temperatures rose (Illustration 4). Is there a relation between these two trends? Would an increase in temperature mean that more irrigation water would be needed than could be delivered by the existing canal system? Are tube wells a response to increased demand? This would seem the logical explanation, but in fact they are not.

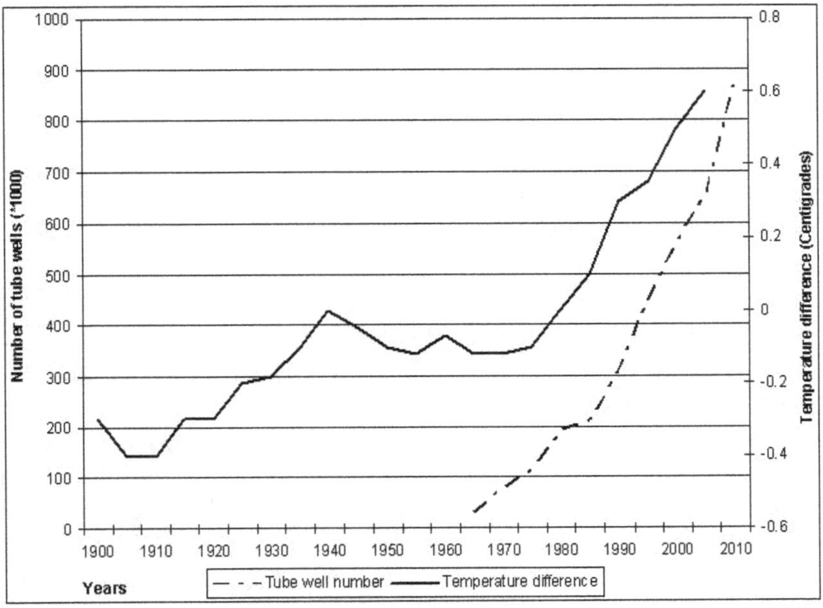

Illustration 4. Trends in temperature change and tube well numbers (temperature Solomon et al (2007); tube wells Qureshi et al (2010))

3. 1800-1950: Irrigation development in the Indus Valley

One could fill many libraries with documents produced on the Indian Colony, and irrigation is not neglected in these documents. British irrigation policies and projects were driven by two considerations that had to be carefully balanced. Irrigation could create prosperity, productivity and political-economic stability. At the same time it was a precarious undertaking requiring continuous attention to technical and managerial questions (Van Halsema 2002: 35). British attention focused on constructing the main canal system; water users had to construct the canals to bring water to the fields. From a policy perspective, the most important question for British colonial administrators was whether land could be irrigated rather than how it was irrigated. They were interested in developing irrigation as a means to increase revenues from irrigable areas. Land taxes were based on the possibility of irrigating the land, not on production increases resulting from irrigation. The British provided Indian agriculturists with water, but what landholders did with it was up to them, as long as they paid taxes.

The British wanted to realize the enormous irrigation potential of the area and lost little time in starting the construction of canal systems. British canal-building activity in India started in 1817 and concentrated largely on the plains to the north of Delhi and the deltaic regions of Madras, South India (Mollinga 1998, Stone 1984). The annexation of the Punjab in 1849 meant that the British gained control over the Indo-Gangetic plain. In the Indus-basin, "huge tracks of crown-waste land could be converted into the granaries of the Raj, by means of the so-called 'colony irrigation systems', that yielded impressive levels of remuneration of up to 45 percent." (Van Halsema 2002; 38). The first large system was the Ganges Canal, opened in 1854, and constructed as a response to a severe famine in 1837-38 (Jurriens et al 1996, Mollinga 1998). For a short period, between 1858 and 1866, the government experimented with irrigation development through private irrigation companies. In 1866 the government decided that canal irrigation should be a state activity (Jurriens et al 1996).

From approximately 1860 onwards, irrigation schemes fulfilled many goals, including famine protection, revenue stability, the settling of unruly tribes, expansion of cultivation, increased production of cash crops, enhanced taxable capacity, improved farming practices, and political stability (Bolding et al 1995). In short, irrigation was an important instrument for almost all fields of governmental policy. The area supplied by irrigation works increased from a little under 2 million hectares in 1878 to almost 4.5 million hectares in 1900 (Jurriens et al 1996; Mollinga 1998). By 1921, some 69 million pounds had been spent on irrigation on almost 18 million acres (about 7.2 million hectares) (Stone 1984).

4. Creating anti-landscapes: from dry to wet

This large colonial gravity irrigation system was introduced in the late nineteenth and early twentieth centuries without a drainage system. As a result, increased irrigation resulted in rising groundwater tables, with associated water logging and salinity problems over large areas (Illustration 5). In order to cope with these problems, the Salinity Control and Reclamation Project (SCARP) was started in 1960. Initially 10,000 tube wells were installed, which not only resulted in the lowering of water table, but also supplemented irrigation. The benefits of the public tube wells motivated farmers to install private tube wells. In order to avoid spending enormous sums on the operation and maintenance of aging public SCARP tube wells, in the early 1980s, the Pakistan Government transferred groundwater development to the private sector by providing financial incentives for tube well development and also transferring some of the existing public wells to private ownership (Johnson 1989). Subsidizing the power supply and introducing small diesel engines also contributed to a dramatic increase in the number of private tube wells. The resulting explosive growth of groundwater exploitation changed the water balance of the Indus basin. The existing irrigation system, with its canal infrastructure and management based on one crop during one of the two seasons, has been modified into a system able to support cropping intensities of 200%.

Since the 1960s, tube well density has increased and acreage per tube well has decreased considerably. Increased tube well density shows that farmers prefer to own tube wells for better control over the timing and quantity of irrigation supply. The development of groundwater through tube wells is not regulated, and anyone who owns a piece of land can drill a well. As reliance on groundwater for crop production is very high and groundwater depletion and quality deterioration have already started, close spacing of tube wells may result in water rights conflicts among farmers. It is clear that groundwater use increases towards the tail ends of canals, due to the rather obvious reason that less canal water is availabile downstream in gravity systems. Irrigation with groundwater of lesser quality has already started processes of secondary salinization. Water tables fall at higher rates towards the tail reaches of canals. Farmers respond to this by increasing the number of deep tube wells. A recent field study found that tube wells with bore depths ranging from 200 to 300 feet had increased from 7% to 30% between 1989 and 2011(Kazmi and Ertsen 2011).

Although strictly speaking, one cannot rule out temperature changes as a partial cause of increased dependence upon irrigation and increased use of ground water, it is clear that the first larger-scale use of tube wells developed in the 1960s in response to changes in groundwater levels caused by the irrigation system put in place during the preceding century. Farmers took advantage of tube wells that were initially intended to lower the groundwater table, to put more water on their fields. They found tube well water supplies much more controllable than canal water ever was. When they saw that tube wells increased both irrigated area and crop production, the Pakistani government began to promote tube wells. I would suggest that the rise in tube-well-fed irrigation in the twentieth century Indus Valley resulted from the SCARP program's response to groundwater rise, which was set in motion by decisions made a century earlier by the British, and from Pakistani public policies that promoted tube wells, more than it resulted as a response to long-term changes in temperature and precipitation.

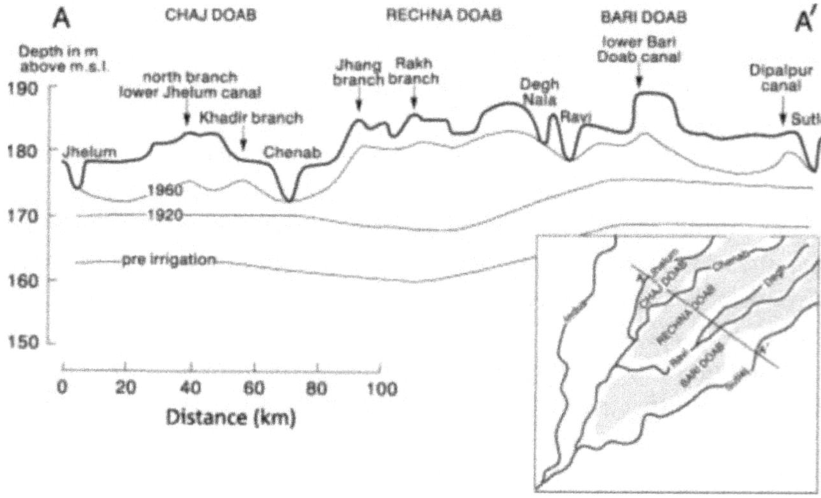

Illustration 5. Indus Basin groundwater trend (Tanji and Kielen 2002)

5. Issues of climate and human agency

Humans create irrigation systems to modify water availability, and thus landscapes, in response to climatic conditions. Irrigation systems may have a considerable impact on the hydrological balance of the watershed in which they are situated, both on the surface water and groundwater. This is especially the case in mountainous environments where groundwater is often near the surface, spatially highly variable, and subject to strong seasonal changes; in less hilly environments, however, particularly when irrigation systems are elements within a larger complex of dams and reservoirs, the totality of man-made structures to manipulate water flows can alter the total water balance significantly, as we have seen in the Indus area. Thus, human agency creates new irrigated landscapes that respond in new ways to climatic contexts (see Carlstein 1982).

The water balance of an irrigation system, including surface and groundwater, cannot be isolated from the watershed within which it exists. Irrigation systems are highly open systems, not only in terms of water flows between the irrigation system(s) and the catchment but also in socio-economic terms. The interaction between water regimes within irrigation systems and the water regime of the surrounding catchments is not yet clearly understood; it is even difficult to anticipate and understand the interactions of water use and management within irrigation systems. Thus, it is not easy to develop a good understanding of water balances in irrigation over time, either in terms of daily flow patterns and seasonal water budgets, which would not only take into account interactions between surface and groundwater, or in terms of the way changing water environments interact with society. Nevertheless, these irrigated landscapes are a continuous process of structuration, in which

> daily routines link to social and material structuration processes on smaller temporal and spatial scales, and … these routines transfer into longer term and large-scale continuities and transformations. […] Irrigation systems develop as emerging anthropogenic landscapes from the purposeful, coordinated but often also uncoordinated activities of individuals, households, and small groups. An irrigation system may be initiated under central rule, by a strong state, but daily practices on the smaller scale would still determine success or failure of irrigation considerably. (Ertsen 2010; 167; see also Sewell 2005)

The farmers' individual and collective agency is the key in understanding irrigation systems. Many farmers worldwide use groundwater pumps, recycle drainage water or construct farm reservoirs to supplement irrigation. All of these mechanisms strengthen farmers' control by freeing them from their dependence on large supply canals that are usually controlled by outside institutions (Ertsen 2009). The Indus Basin case shows the complexity of cause and effect of observed developments. Can we attribute the tube well growth to the colonial British irrigation policy? Obviously not, but there is a relation between the two because SCARP built the first public tube wells in this region in response to the effects of colonial irrigation on ground water. This, in turn, created new conditions and opportunities for farmers and the Pakistani government. Is climate to be ruled out completely? Probably not. A cause-effect analysis on the effect of climate on human activity is difficult, but rising temperatures and possibly changing rainfall would have an influence on crop water demand and influence water use.

Irrigation shows the extreme complexity of, as well as the complex relations between, human actions and their social and material contexts. Irrigation is shaped through complex interactions between humans and environment (see Ertsen 2010; Janssen and Anderies 2007; Janssen et al 2006; Kohler and Van der Leeuw 2007; Marcus and Stanish 2005; Nelson et al 2010; Peeples et al 2006; Scheffer et al 2009). The natural environment sets boundary conditions for humans, to which they respond with irrigation technology, with such responses impacting both boundary conditions and humans anew. Because of this complexity, it is extremely difficult – but hopefully not impossible – to develop better understanding of the spatial and temporal scales of these interactions, and of changes to irrigation systems, the societies that build these systems, and the larger hydrological and climatic regimes in which they operate. Not only do we need a better understanding of the effect of material conditions on human actions, but we also need greater comprehension of the connections among individual decisions, rules and past actions within irrigation systems. Recognizing human-environment interactions as historically contingent, shaped by distinct cultural perceptions and past human actions, is vital. Landscapes are not stable, nor are irrigation systems.

6. Creation as a first act of destruction

In this essay I have sketched contours of an interdisciplinary effort combining archaeology, history and human geography. This brief outline is part of larger research into water availability and allocation in (ancient) communities and their dependent connection to engineered landscapes – not restricting engineering to the modern concept. Knowing the quantitative and qualitative dimensions of water access by an ancient population reveals important aspects of the kind and degree of socioeconomic and socio-political organization in the human past.

> [...] the systems with the best chance for uninterrupted longevity have slowly evolved on the highly variable landscapes from which people make a living. Even under appreciable stress, water management systems tend to persevere because of their adaptability. This aspect of water management receives less attention because it is less spectacular than the origin or collapse of a system. Nevertheless, societal maintenance and sustainability deserve greater scrutiny in our rapidly changing world. (Scarborough 2003; p 3-4; see also Scarborough and Isaac 1993).

This is not to say that water manipulation was the single most significant factor in stimulating social development and complexity; such an idea has been shown to be highly reductionist if not misleading. Nevertheless, water remains the most primary of resources on our planet and the most precious of all resources that humans can control – more so than even food. It requires major efforts to command and significant infrastructural investment to accommodate population densities and growth. If we can articulate how past irrigation systems were organized and used, we may be able to evaluate aspects like control, resource access and the economic organization of their societal institutions.

Archaeological case studies may provide the *longue durée* necessary to properly assess the robustness of irrigation systems and societies. Learning how resilient environmental infrastructures and investments may or may not have been in the past, and establishing parameters that could indicate societal success – the latter something that the archaeological record already provides – will contribute to the growing empirical literature concerning how we can best incorporate and sustain our planet's limited and diminishing resource base. Viewed through a prehistoric or historic lens, the differences and similarities between irrigated landscapes, anti-landscapes, and the societies that created them may offer better understanding of current issues (Hughes 2004). Clearly, present problems have historical roots, (Tvedt 2004), and the Indus farmer of today is connected to the British irrigation engineer of a century ago.

Bibliography

Abbott, D. 2003. *Centuries of decline during the Hohokam Classic Period at Pueblo Grande*. Tucson: University of Arizona Press.

Agricultural Statistics of Pakistan, 2008-09. Ministry of Food and Agriculture, Economics Wing. Government of Pakistan.

Akkermans, P.M.M.G., Van der Plicht, J., Nieuwenhuyse, O.P., Russell, A., Kaneda, A. and Buitenhuis, H. 2010. Weathering climate change in the Near East: dating and Neolithic adaptations 8200 years ago. In: *Antiquity* 84(325).

Aslam, M., Prathapar, S. A. 2006. *Strategies to mitigate secondary salinization in the Indus Basin of Pakistan: A selective review*. Research Report 97. Colombo, Sri Lanka: International Water Management Institute (IWMI).

Bolding, A., Mollinga, P.P. and Van Straaten, K. 1995. Modules for modernisation: colonial irrigation in India and the technological dimension of agrarian change. In: *Journal of Development Studies*, 31(6): 805-844.

Carlstein, T. 1982. *Time resources, society and ecology. On the capacity for human interaction in space and time, Volume 1: pre-industrial societies*. London: Allen and Unwin.

Cheng, H., Fleitmann, D., Edwards, R.L., Wang, X., Cruz, F.W., Auler, A.S., Mangini, A., Wang, Y., Kong, X., Burns, S.J. and Matter, A. 2009. Timing and structure of the 8.2 kyr B.P. event inferred from $\delta^{18}0$ records of stalagmites from China, Oman, and Brazil. In: *Geology* 37: 1007-1010.

Costanza, R., Graumlich, L.J. and Steffen, W. 2007. *Sustainability or collapse? An integrated history and future of people on earth*. Cambridge: MIT Press.

Diamond, J. 2005. *Collapse: how societies choose to fail or succeed*. London: Penguin.

Drysdale, R., Zanchetta, G., Hellstrom, J., Maas, R., Fallick, A., Pickett, M., Cartwright, I. and Piccini, L. 2006. Late Holocene drought responsible for the collapse of Old World civilizations is recorded in an Italian cave flowstone. In: *Geology* 34: 101-104.

Ertsen, M.W. 2010. Structuring properties of irrigation systems. Understanding relations between humans and hydraulics through modeling. In: *Water History* 2(2): 165-183.

Ertsen, M.W. 2009. From central control to service delivery? Reflections on irrigation management and expertise. In: *Irrigation and Drainage* 58: S87-S103.

Fisher, C.T., Hill, B.T. and Feinman, G.M. 2009. *The archaeology of environmental change. Socionatural legacies of degradation and resilience*. Tucson: University of Arizona Press.

Hijma, M.P. and Cohen, K.M. 2010. Timing and magnitude of the sea-level jump preluding the 8200 yr event. In: *Geology* 38: 275-278.

Howard, J.B. 1993. A paleohydraulic approach to examining agricultural intensification in Hohokam irrigation systems, in: Scarborough and Isaac (eds.) (1993): 262-324.

Hughes, T.P. 2004. *Human-built world: how to think about technology and culture*. Chicago: University of Chicago Press.

Janssen, M.A. and Anderies, J.M. 2007. Stylized models to analyze robustness of irrigation systems. In: Kohler and Van der Leeuw (2007): 157-173.

Janssen, M.A., Bodin, O., Anderies, J.M., Elmqvist, Th., Ernstson, H., McAllister, R.R.J., Olsson, P. and Ryan, P. 2006. Toward a network perspective of the study of resilience in social-ecological systems. In: *Ecology and Society* 11(1): 15.

Johnson, R. 1989. *Private tube well development in Pakistan's Punjab Review of past public programs/policies and relevant research.* Colombo, Sri Lanka: 36p.

Jurriens, M., Mollinga, P. and Wester, P. 1996. *Scarcity by design. Protective irrigation in India and Pakistan.* Liquid Gold Paper 1. Wageningen: Wageningen University.

Kazmi, S.I and Ertsen, M.W. 2011. Changes in tube wells count in two decades in area under conjunctive irrigation in Pakistan. Paper presented at the *25th ICID European Regional Conference* (Groningen, May 16-20 2011).

Kohler, T.A. and Van der Leeuw, S.E. (eds.) 2007. *The model-based archaeology of socionatural systems.* Santa Fe: School for Advanced Research Press.

Marcus, J. and Stanish C. (eds.) 2005. *Agricultural Strategies.* Los Angeles: Cotsen Institute of Archaeology.

McAnany, P.A. and Yoffee, N. 2010. *Questioning Collapse: Human resilience, ecological vulnerability and the aftermath of empire.* Cambridge: Cambridge University Press.

Mollinga, P.P. 1998. *On the waterfront. Water distribution, technology and agrarian change in a South Indian canal irrigation system*, PhD thesis Wageningen University.

Nelson, M.C., Kintigh, K., Abbott, D.R. and Anderies, J.M. 2010. The cross-scale interplay between social and biophysical context and the vulnerability of irrigation-dependent societies: archaeology's long-term perspective. In: *Ecology and Society* 15(3): 31.

Peeples, M.A., Barton, C.M. and Schmich, S. 2006. Resilience lost: intersecting land use and landscape dynamics in the prehistoric southwestern United States. In: *Ecology and Society* 11(2): 22.

Qureshi, A.S., McCornick, P.G., Sarwar, A. and Sharma, B.R. 2010. Challenges and prospects of sustainable groundwater management in the Indus Basin, Pakistan. In: *Water Resources Management* 24: 1551-1569.

Scarborough, V.L. 2003. *The flow of power. Ancient water systems and landscape.* Santa Fe: SAR Press.

Scarborough, V.L. and Isaac, B. (eds.) 1993. *Economic aspects of water management in the prehispanic new world.* Research in Economic Anthropology, Supplement 7, Greenwich.

Scheffer, M., Bascompte, J., Brock, W.A., Brovkin, V., Carpenter, S.R., Dakos, V., Held, H., Van Nes, E.H., Rietkerk, M. and Sugihara, G. 2009. Early-warning signals for critical transitions. In: *Nature* 361(3): 53-59.

Schwartz, G.M. and Nichols, J.J. 2006. *After collapse: the regeneration of complex societies.* Tucson: University of Arizona Press.

Sewell, W.H. 2005. *Logics of history. Social theory and social transformation.* Chicago: University of Chicago Press.

Solomon, S., Qin D., Manning M., Chen Z., Marquis M., Avery K.B., Tignor M. and Miller H.L. (eds.) 2007. *Contribution of Working Group I to the Fourth Assessment Report of the Intergovernmental Panel on Climate Change.* Cambridge: Cambridge University Press.

Staubwasser, M. and Weiss, H. 2006. Holocene climate and cultural evolution in late prehistoric-early historic west Asia. In: *Quaternary Research* 66: 372-387

Stone, R. 2006. The end of Angkor. In: *Science* 311: 1364-1368.

Stone, I. 1984. *Canal irrigation in British India: perspectives on technological change in a peasant economy.* Cambridge: Cambridge University Press.

Tanji, K. and Kielen, N. 2002. *Agricultural drainage water management in arid and semi-arid areas*. Food and Agricultural Organization, Irrigation and Drainage Paper 61.

Tvedt, T. 2004. *The River Nile in the age of the British. Political ecology and the quest for economic power*. London: I.B. Taurus.

Ullah, M. K., Habib, Z. and Muhammad, S. 2001. *Spatial distribution of reference and potential evapotranspiration across the Indus Basin Irrigation Systems*. Colombo: International Water Management Institute, Working Paper 24.

Van Halsema, G.E. 2002. *Trial and re-trial; the evolution of irrigation modernization in NWFP, Pakistan*. Dissertation Wageningen University.

Weiss, H., Courty, M.-A., Wetterstrom, W., Guichard, F., Senior, L., Meadow, R. and Curnow, A. 1993. The genesis and collapse of third millennium north Mesopotamian civilization. Science 261(5124): 995-1004.

Wilkinson, T.J. 2003. *Archaeological landscapes of the Near East*. Tucson: University of Arizona Press.

4

Landscape, Anti-landscape and the Western Political Imagination: J.B. Jackson's Challenge to Environmentalism

Mark Luccarelli

J.B. Jackson's writings on actually existing American landscapes reveal their political and historical qualities, which he sharply contrasted with romantic and post-romantic conceptions of the environmental landscape. This essay examines the implication of Jackson's work for environmental studies, suggesting that the landscape/anti-landscape binary has broad symbolic and political significance and may be a useful starting point for re-contextualizing questions of environmental governance. Landscape, anti-landscape, the western political imagination, J.B. Jackson, environmentalism

Keywords: Anti-landscape; environmentalism; Jackson, J.B.; landscape; Lefebvre, Henri: Marx; Leo; Poussin; Nicolas; Pueblo Indians

1. Introduction

In announcing the symposium on which this volume is based, David Nye defined the anti-landscape as

> … spaces that do not sustain life. They may arise from natural causes but are often the unintended consequence of human action. This is a call for interdisciplinary papers that deal with either the emergence or the recovery/salvaging of such spaces … The concept of the anti-landscape provides a point of unity for work that deals with sustainability, dystopia, environmental degradation, resilience, "the tragedy of the commons," and the overuse of resources.

Clearly anti-landscape considered as a dysfunctional space that fails to support life has helped to define its opposite: a healthy, sustainable landscape. There is much evidence to support this as a theme in environmental studies. From nineteenth century conservation to ecological and post-ecological discourses, writers have observed and interpreted the anti-landscape. Nineteenth and early twentieth century resource conservation in the United States was helped immeasurably by George Perkin Marsh's 1864 book, *Man and Nature*. Marsh describes the anthropogenic "physical revolutions" that have destroyed forests and altered the face of the planet, damaging soils, and altering the carrying capacity of the land and perhaps even altering climates (Marsh, 1864/2003: 18). A more dramatic anti-landscape marked by chemical or nuclear poisoning emerged to public notice with Rachel Carson's *Silent Spring*, published in 1962, which established the foundation for what Lawrence Buell calls the "toxic discourse" (Buell, 2001). In an article on *The Maine Woods*, I argue that Thoreau's idea of wilderness was developed in relation to a feral anti-landscape – lightly abused but devalued woodlands left after the (temporary) cessation of limited logging operations (Luccarelli, 2010). Recent environmental studies' interest in the anti-landscape marks a worthy effort to reject social constructivist understandings of environment as epiphenomenal (Buell, 2001: 31), on the one hand, and at the same time to find a spatial framework and concrete political context for the environmental imagination.

The discourse has not been without its problems, however. Buell argues that "toxic discourse," illustrated by the story of a doomed suburban neighborhood along the Love Canal, broke down the discrete city versus country categories established by pastoral nature writing and the wilderness romance, creating an "'ecology' of place," that "democratizes the pastoral ideal" by presenting the right of all to "clean, uncluttered surroundings" (2001: 38, 46). But the discourse raises as many questions as it addresses and provides us with little ammunition to answer the question of what form a clean, uncluttered "place for living" in a post-urban age should take. Indeed by the 1990s the popular discourse on urban "sprawl" as illustrated by books, such as James Kunstler's *The Geography of Nowhere*, have made Leo Marx's critique of the "simple pastoral" – the self-defeating and ever accelerating search for clean, uncluttered countryside – a commonplace. At the same time, the oft-expressed theme of "return to the city" is not always a financially or socially viable alternative for the middle and working classes throughout the western world (Kunstler, 1994; Marx, 1964).

A larger and arguably more difficult problem is how to represent "environment" in light of contemporary sciences' far more sophisticated, but also far more confusing, understanding of anthropogenic effects on environment. Anti-landscapes like those generated in hurricane zones of the American Gulf Coast or in flooded Arctic icescapes, can no longer be clearly understood as either "natural" or "human" centered. The blurring of what had been clear demarcation between anthropogenic and geo-biological spheres has posed a threat to the capacity to generate visions and programs of environmental reform; indeed the underlying meaning of the word "environmental" has been undermined. Clearly in our "post-modern" era, we are witness to the rise of the anti-landscapes as post-landscapes, i.e. a sphere that cannot be visually imagined and even cannot even be identified as a distinct and discrete part of the earth (Garrard, 2011). The loss of territoriality in environment comports poorly with the overwhelming territoriality of human experience. Within a single generation, relatively stable modes of representation drawn from the Humanities – literature, art and architecture, photography – have been supplanted by mapping, satellite imagery and blogging – modes of representation inherently unstable and distant from their referent. With "environment" being increasingly defined as borderless and global, we get the sense that events are outside of our control: little wonder that pre- or post-apocalyptic views of life packaged by the video-and-game sphere have become commonplace. There are differences among environmentalists on the question of electronically-mediated culture and its capacity or lack thereof to express alternatives, but all environmentalist thought assumes that representation is the fundamental question, which leaves it vulnerable to the current problems of representation in a realm of destabilized electronically mediated culture. All assume then that environment – our surroundings – can and will speak directly to us, if adequately and forcefully represented. We must look carefully at this assumption.

2. Jackson's Critique: Landscape as Political and Lived

John Brinckerhoff Jackson (1909-1996) was an American writer, artist, critic and publisher. Born in France of American parents and educated in Europe and the United States, he attended the University of Wisconsin and Harvard but never took a degree. After enlisting in the U.S. Army in 1940, he was stationed in Europe and worked as a translator during much of the war; there he became in the interested in the work of the *Annales* School of historical geography, which helped to formalize a lifelong interest in landscape study. Returning to the United States, he founded the magazine *Landscape* in 1951 and lived much of his life on a ranch just outside of Sante Fe, New Mexico (Oakes & Price, 2008: 153-154).

Jackson was part of a generation of public intellectuals that includes Van Wyck Brooks and Lewis Mumford – intellectuals and essayists, who operated outside of the academy and were also important in American Studies during its first national phase. In his mature work, published during the 1980s and 1990s, Jackson considered such themes as the rise of environmentalism, urban and landscape preservation and the character of academic environmental discourses. His perspective is interesting because Jackson criticized, in fact fundamentally disagreed with, the rise of an environmentally centered concept of landscape as historically inaccurate and politically misplaced. While I will argue that Jackson misses an important part of the larger picture necessary to the re-grounding of environmental studies in a more sophisticated view of space, his critique provides an important and convincing exposé of a central problem of environmental discourse.

I return to the question of the anti-landscape in a moment. To begin with it is useful to examine the use of the term of landscape. In two essays entitled, "The Word Itself" and "A Pair of Ideal Landscapes," that form the theoretical introduction to his 1986 book, *Discovering the Vernacular Landscape*, Jackson takes issue with the rise of environmental studies, which has, as he sees it, appropriated the word landscape and interpreted it as a "natural feature of the environment" (1986a: 8). There is no distinction implied here between scientific and humanistic environmental studies. The development of an environmentally centered conception of landscape could be understood as one part of a broadening of the word "landscape." What had been a very specific term applied to art and architecture concerned specifically with the achievement of aesthetic effects as defined by artists and critics was returned as Jackson puts it, to "the public domain" where it was open to speculation (1986a: 4).

Jackson doesn't object to the loss of its nineteenth century meaning, in part because, like the environmental discourses he criticizes, he became less interested in the conventional romantic visual landscape himself.

Actually in his opening essay, "The Word Itself," Jackson's primary point is to argue that everyday landscape is largely invisible. A "composition of man-made spaces on the land," landscape blends seamlessly into the natural environment so as to be virtually undetectable. This is less true of topographical reorganizations of space such as are effected by building dams or irrigation systems, but for the most part landscape is "produced by simple spatial reorganization" (1986a: 7-8). We can see this etymologically: "land" signifies an enclosed space and a "scape" meant a man-made sphere.

For Jackson the etymology of the word reflects practice – the ingrained habits and patterns of inhabitation, of the conceptualization and relation to land over deep time in addition to the political organization of space by nation-states (to which we might add the spaces of transnational organization). The point is not only that landscape reflects the primary ways in which we have conceived and related to inhabitation of the earth's surface, but that landscape is really largely a reflection of the social world, both in its boundary-making and use-dependent aspects.

Landscape is a reflection, then, of a community, both of local communities engaged in their vernacular traditions and habits of occupying the land, and modern political communities with their boundaries, highways, monuments and public places.[1] In "A Pair of Ideal Landscapes," the vernacular or "inhabited" landscape – the spatial practices originating in the occupation of the land by folk cultures/informal communities – interested Jackson, especially in regard to the attributes of fluidity and mobility in occupation of the land, and he asserted their continued relevance in the making of contemporary vernacular landscape, in North America in particular. But the "traditional inhabited landscape" arrives not from design but through centuries of practice and arrives at some sort of pattern only after "generations of unrest and confusion." The political landscape, by contrast, is the "realization of an archetype" (Jackson, 1986b: 43). Jackson asserts boundaries are perhaps the most important feature of the political landscape and he asserts that in many respects the

1. Kenneth Olwig (2002: 25-26) tells us that *landschaft* in 15th century Germany was the organization of community life that embodied obligations for maintaining roads and bridges and establishing practices for cultivating fields – part of what Jackson calls the landscape of inhabitation: the foundation of the vernacular. See also Powell (1963) & Bloch (1966).

act of boundary-making has very little to do with the physical environs or even, in the more extreme, cases with the social environment:

> A nation, we say, is not simply a collection of people, it is also the territory they occupy, and the boundary should be drawn so that the two entities correspond as closely as possible. … The traditional political landscape had a very different conception of the boundary: it was intended to less define a region and establish an effective relationship with the outside world than to isolate and protect something within it. … A good illustration of this can be seen in the United States … (it) created a number of states which were in effect immense rectangular spaces having no relation to topography or population (1986b: 14).

Like the classical republican city-state, the purpose of American state boundaries was to isolate and protect people and to create durable political forms within those protective boundaries. In many respects the political landscape separates us from environment by imposing public purposes on space. The American system of dividing land into rectangular sections accounts for a major feature of landscape in approximately 2/3 of the United States: it completely ignores environment, both in terms of topography and the native fauna and flora. Its purpose was to secure tenure for a class of small holders meant to be the foundation of the American republic.

Granted, these highly purposeful impositions of a political landscape have eroded over the past two generations. Jackson observes that classical public space, linked visually to the Italian piazza or the Spanish plaza, took the form of the American courthouse square of the early republic and was restated in the turn-of-the-twentieth century "civic centers" of the City Beautiful movement; but he adds that the political-social purpose of these public spaces had been lost in the society of the 1980s to a new quasi-public space as a sphere of "passive enjoyment" – an entirely passive "environmental" conception of space, which cannot generate principles sufficient to reshaping the urban landscape (1986b: 17).[2]

The global aspirations of today's successful nation-states have imposed another layer on the political landscape, but even this change doesn't challenge Jackson's point: environment is over-ridden by both politically derived and ideologically sanctioned patterns and by long term usage of the land. Jackson's second main point is that environ-

---

2. As early as 1959, Jackson tells us that "to the average man the immediate experience of Grand Central [Terminal in New York] is neither architectural nor social: it is sensory." But he seems less problematic then; his view evolves over the course of two decades (1959/1970: 83).

mentalism encourages thinking of landscape as a response to the natural or ecological order, in effect displacing civic spaces with natural ones. A strong environmental discourse simply confuses environmental and public purposes. They are and can be discursively related, of course, but this takes a new discourse that would establish the necessary linkages. At the same time despite the confusing spatial sphere created by new highways, lifestyle changes and consumer culture, people still manage to generate new public spaces, but no longer by design. They are made apart from the earlier classical contexts, and are now found in school auditoria, college campuses and shopping malls – and I should add in virtual spaces. Perhaps more importantly, other aspects of political space retain vitality. Jackson mentions roads as perhaps the second most important element of the political landscape (after boundaries). Comparing ancient Rome to the USA, Jackson sees road networks vast in scale, created without reference to topography and built to strengthen commercial and military functions of the society. Roads are the antipathy to place awareness because they remind us "man as a political animal is always inclined to be footloose, inclined to leave family and home for a more stimulating place" (1986b: 27).

We can conclude first that the historic political landscape and the traditional inhabited landscape are fundamentally at odds with the environmentally defined landscape, and second that in many respects the political definition of space forms the fundamental guide by which a polity, any polity, inhabits the land. What the political and inhabited landscapes have in common, as against environmental conceptions of space arrived at through simulation, semiotics or phenomenology, is *visibility* recognizable to the layman (1986b: 43). Where does this leave the hopes for advancing environmental goals through the reformation of landscape? The problem concerned Jackson as well. In the early 1990s he wrote an essay entitled "In Search of the Proto-Landscape": a "basic landscape from which all existing landscapes are mere variations." By "proto-landscape" I take Jackson to mean an ideal type through which all other landscapes – including the vernacular landscape of north-western Europe and the political landscape of the Classical world[3] – may be measured, rather than an historically precedent landscape from which all others are said to derive. Perhaps a proto-landscape might illustrate principles from which we could derive

3. The American landscape is derived from these two sources.

a more balanced relation to the natural world? Jackson turned to a nearby source, the ancient Anasazi culture and their contemporary "Pueblo" descendants of the American southwest.

Landscape for the Pueblo people, he finds, was structured around "place, location and direction" and differs dramatically from the flexible and fluid idea of landscape in the Euro-American cultural field. The Pueblo landscape reflects, in part, the geo-physical clarity of the desert southwest, its natural space, but it also reflects on their *mentalité* as well. The Anasazi landscape is conceived in four concentric zones: the village itself, the surrounding sphere of irrigated gardens and fields, a zone of grasslands or rangelands and finally a frontier zone marked by high and inaccessible mountains. In the Pueblo world, center and periphery are the fundamental concepts. The center is the village (pueblo) and at its center was (is) the village plaza, which is conceived as "Earth Navel" of "Middle Place" – the holiest of shrines. The periphery is a container that gives structure to place at the center. The boundary or "frontier zone," marked by four mountains, was a kind of "container of psychic power" that assured the triumph of order over chaos and that was linked mythically (through narrative) to the ritual center of community life (1995: 45). Jackson also points out that the container or boundary line that separates landscape from natural environment that lies outside it was a feature of the traditional Germanic concept of the *mark* (the *markstein* in German means the stone delineating boundaries). Organizing landscape around concentric zones is not uncommon and Jackson points out a similar structure in medieval Europe.

Nonetheless, Jackson is compelled to admit that though they share the quality of being a political-ideological construct related though not reduced to actual practices of inhabitation, ancient Amerindian/European medieval landscapes and modern landscapes differ in fundamental ways. Far from a universal, the construction of landscape follows ideal types in which different aspects of landscape are variously expressed: medieval and ancient landscapes are less political and more environmental in their construction, while ancient Greece exemplifies a political execution of landscape.[4]

4. A contrary view is that response to environment always structures the lifeworld and therefore perspective. For example, "the cosmos of the Indians is spatially well-defined, stratified and rotary" which comports with a "layered" landscape marked by sharp contrasts (Tuan 1974/1990: 81).

Some of these understandings are more compatible with an ecological understanding than others (Jackson, 1995: 44). In ancient cultures a religious cosmology is the source of ultimate value; religion trumps the political landscape and the natural world is given a decided place in the order. In a modern landscape that values mobility and freedom above all else, environment has a largely instrumental value. Regarding the academic debate between naturism and social constructivism (Buell 1996), Jackson would have surely agreed about the reality of the natural environment, but would have disagreed that landscape can be adequately represented through a description of the natural environment.

This question aside for a moment, the advantage that naturism has is its consistency: it achieves the coherency made possible by a unified philosophical perspective, while social constructivism suffers all the confusions and multiplicities of modern society – all of which are necessary to our political imaginary and have made it ungovernable. For these reasons Jackson was skeptical about the prospect of environmental reform – a skepticism which has been more than warranted by the failure over the approximately two decades since his death in making progress in this field. Discontinuity and chaos seem to be the legacy of a classical political landscape foundering on the rocks of commercialism and political ennui. Jackson concludes in a moment of resignation to *fortuna* that only with the passage of time and through the attainment of a long-term occupation of the land will "we will learn to perceive … (commonplace) spaces as symbols of continuity and order …" such that we see "in them landscape beauty of the elemental kind that transcends history" (1986b: 35).

In effect, Jackson's work says to us that however rhetorically sophisticated we are in our technologies of gathering and representing the surroundings, and regardless of how strong an environmental consciousness we develop, we will never address the problem of how to occupy the land in a sustainable way without formulating a "moral-practical end" attached to an "instrumental means."[5] Environmental discourses have identified moral ends, but not in practice and certainly not in relation to the instruments – technological and political – through which alternatives

5. The terms are Paul Goodman's (1962: 16): "For Dewey and for James, the end and the means, the moral-practical and the instrumental, derive meaning from their intrinsic connection in process, and there is no other (source of) meaning."

may be constructed. The neo-classical political landscape that Jackson identified was itself a powerful instrument reflecting the moral-practical ends of Jeffersonian democracy. Through this instrument environment was governed to certain specified ends. One may argue that the instrument has broken down, is in a state of decay, or has been complicated by the layering of other political landscapes on top of the old. Fair enough, but the point remains that to construct alternatives requires both new instruments and re-stated/re-defined ends: a reconstructed political landscape has the potential to encompass environmental ends. Thus governance of environment is not "environmental," i.e. lodged in the response to and consciousness of the world, or in the adjustment to natural systems: governance is the re-ordering of the political and lived landscapes that Jackson identified. But does this mean that environmental awareness is useless? Actually not: what is required is an environmental understanding that considers ways that environmental goals have been inscribed onto the political landscape, in effect establishing a relation between environmental awareness and the political imagination, but in a way that keeps alternative formations foremost in mind.

3. From Jackson to Lefebvre: landscape/anti-landscape binary as an expression of the unfolding of space

Thus far I have relied on Jackson to argue that the actually existing landscape cannot be understood in environmental terms, that it is quite fundamentally a system for the political and social organization of space. Jackson identifies the (often) invisible quality of landscape to organize human life and the natural world into spatial system, which blends political and practical meaning with aspects of environmentality. Now I wish to go beyond Jackson to suggest that the landscape/anti-landscape binary may be understood as the expression of a conflict over the formulation and reformulation of the environmental attributes of space. Beneath the representations of landscape is a binary driven deeply into the western political imagination between order and freedom in which various conceptions of environment have played a role.

For Jackson the possibility of engagement or of a dialectic of alternative conceptions of environment was lost in the erosion of the American political landscape – a loss which could find its only potential compensation in a deep time process by which a new vernacular arises through evolving patterns of land use. Jackson emphasized the historical over the environmental/aesthetic and the political as against the social origins of landscape (1986b: 18, 33). Henri Lefebvre, by contrast, returns to the earliest beginnings of modernity, to the first great political landscape of western modernity, European urbanism, to show how landscape as a social construction and a visual and symbolic expression has always contained its own antithesis. This conclusion in turn rests on the tensions in space-making as a tri-parte process encompassing the "abstract space" of capital, the perceived space that arises from the everyday lifeworld, and the "spaces of representation" from which the landscape imaginary arises (1974/1991). Denis Cosgrove takes spaces of representation to mean that landscape reflects a *symbolic order* engaged with, and not an epiphenomenon of, the social and political order of its day. It is a symbolism that draws on an environmental sensibility or "environmentality" – in the broad sense of reflecting the social and political landscapes, the built environment with its divisions and organizations of space in relation to nature; therefore landscape symbolically defines the beautiful and the true, reconciling aesthetics with social dynamics and society with (a current understanding of) nature. Thus one dynamic of space-making relevant to the question of the significance the landscape/anti-landscape binary reflects the sense that symbolic landscape is socially-constructed to encompass environment into the reigning spatial order, while anti-landscape discovers an aesthetic (and often moral) principle of dissonance based on an observation of landscapes that resist assimilation into space-making. Established uses of environment and senses of environmentality are violated by anti-landscapes, whether they derive from the repossession by the writer of classical pastoral or from a reading of contemporary experience, as for example in an aesthetic that attempts to capture a particular life-style or (in Lefebvre's typology) the dynamics of everyday lived space. In the process, the anti-landscape gives expression to a new sense of environment that better represents human aspiration and understanding of nature. The other side to the process of anti-landscape formation, however, is the capability of any given spatial order to assimilate and consolidate an anti-landscape into the reigning symbolic order.

Historically, the greatest achievement of the imaging of a symbolic landscape was the European city predicated on the creative use of geometric space, the careful consideration of architecture in relation to setting and the decorative function of public art. Denis Cosgrove describes the resulting "idea of landscape" as the capacity to reflect what is seen with socially influenced ways of seeing for both mainstream and alternate cultural formations (1984/1998).

Illustration 1. Anonymous (attributed to Piero della Francesca, Italian, 1416-1492 and others), The Ideal City, ca. 1470, Oil on panel, 60 x 200 cm, Galleria Nazionalle delle Marche, Urbino, Italy; Wikipedia Free Commons.

If landscape takes the form of the geometric city, the Renaissance ideal of mathematical order and geometric perfection, the simplicity and sincerity of which was later lost to the rich décor and elaboration of baroque classicism, then anti-landscapes called what became the apparently seamless and stale classical aesthetic of city-landscape into question. By the seventeenth century, an anti-landscape embodies psychological terror in Poussin's rendition of a man fleeing a storm that casts its shadow over the unstable (though picturesque) setting at the margins of the new city; in the foreground shadows partly conceal a mythological narrative centered on murder. This anti-landscape does not reflect environment apart; it grounds environment in the classical political imaginary torn between order and freedom; and at the same time, the painting is responding to a new experience of the environment – an era of exploration and expansion north and west which would greatly expand the received sensibility of nature as expressed in concepts of wilderness in culture and naturalism in painting.[6] Poussin has managed a new symbolic landscape marked by

6. A view early expressed by Ruskin (1859: 173): "for the most part a great peacefulness of light, Syria and Greece, Italy and Spain, laid like pieces of golden pavement into the sea-blue, chased … with terraced gardens, and flowers heavy with frankincense …" may be contrasted with the "dark forests of the Danube and Carpathians … and then, farther north still, to see the earth heave into mighty masses of leaden rock and heathy moor, bordering with a broad waste of gloomy purple … belt of field and wood…"

the plentitude of natural space which is achieved by his large of depth of field.[7] Poussin's contrast between civilization and nature – and the triumph of the plentitude of natural space – points to the origin of Eco-literatures and naturalistic landscape painting as well. As reflective of its age, Poussin's naturism was not "environmentalist" in the sense we currently understand, but it was an expression of representing environment in a new way that better responded to observed landscapes and enlarged human aspirations.

Illustration 2. Nicolas Poussin (French, 1594-1665), Landscape with a Storm, 1651, Oil on canvas, 100 x 130 cm, Musée des Beaux-Arts, Rouen, France; Wikipedia Free Commons.

The decay of order and the emergence of the psychic disorder of the unconscious mind is in environmental studies seen in the early emergence of sublime art, which wrested environmental meaning – a sense of the magnitude of space that must be read in response to a rationalized and

7. "From the outset, Poussin was as concerned with landscape settings as with narrative subjects": (Christiansen, Wrightsman & Rosenberg, 2008).

desensitized spatial order (Butterfield, 2008). The reigning environmentalist narrative holds, by contrast, that while environment became first aestheticized in terms of grandeur and awe (vastness became the sublime), the environmental imagination was distracted by the social and political uses to which it was put. Only when the allied discourses of the genius of national place and "outback nationalism," (Buell, 1996) were left behind, could a true environmentalism aimed at saving nature develop. Contemporary environmental discourses, having exposed these political deviations from its message of increasing awareness of the environment, is now poised to push for a deeper understanding of the meaning of inhabitation based on reconciliation with the natural environment.

Of course attention to the discourses of limits and toxicity is important, but my view is that acts of the environmental imagination are always embedded in political and social perspectives. Seeing landscapes in symbolic terms, they will always obscure as much as they reveal. They reveal the possibility of alternative spatial formations; they reveal the terms that must be addressed insofar as they respond to the processes of landscape formation. But they at least partly conceal the means of addressing these issues because, as Lefebvre has it, the third of his spatial triad, lived space, requires a social-political analysis that sees in human practices of occupying space the opportunity to create alternative spatial formations. Spatial formations that respond to the social and political divisions of space are not environmental in this restrictive sense. They raise the broader question of how space making should be governed in relation to how we discover, represent, understand and *re-form* space.

To return to and re-contextualize J. B. Jackson's landscape concept: vernacular should be understood as a commonplace political landscape, rooted in lived spaces. The real significance of such landscapes is that they can generate alternative spatial formations that have the potential to address the structural impasses between symbolic order and freedom, between traditional spatial forms and technological innovations and between human and natural perspectives implied in the landscape/anti-landscape binary. Reading the history of environmental thought through this framework requires an understanding of the political and vernacular foundations of landscape particular to national and regional cultures, and poses a challenge to the direction that environmental discourse has undertaken. I have argued that too often environmentalism has evaded the political – and in particular, evaded the socially and politically derived foundations of space making. Jackson's wide-ranging discourse on landscape is an important part of a fundamental corrective to this perspective.

Bibliography

Bloch, M. trans. J. Sondheimer. 1966. *French Rural Society: An Essay on Its Basic Characteristics*. London: Routledge & Kegan Paul.

Buell, L. 1996. *The Environmental Imagination*. Cambridge, MA: Harvard University Press.

Buell, L. 2001. *Writing for an Endangered World*. Cambridge, MA: Harvard University Press.

Butterfield, A. 2008. "The Magical Painting of Poussin," Review of Poussin and Nature: Arcadian Visions , Catalog of the Exhibition at Metropolitan Museum of Art, New York, ed. by P. Rosenberg & K. Christiansen. New Haven, Conn.: Metropolitan Museum of Art/Yale University Press. New York Review of Books.

Christiansen, K., Wrightsman, J. & Rosenberg, P., curators. 2008. "Poussin and Nature: Arcadian Visions," The Metropolitan Museum of Art, New York, New York, Special Exhibition Galleries: February 12-May 11, 2008. http://arttattler.com/archivepoussin-nature.html, accessed May 25, 2012.

Cosgrove, D. 1984/1998. *Social Formation and Symbolic Landscape*. Madison: University of Wisconsin Press.

Garrard, G. & Hartman, S. 2011. Unpublished interview on the theme anti-landscape.

Goodman, P. 1962. "Utopian Thinking," in *Utopian Essay and Practical Proposals*. New York: Vintage Books.

Jackson, J.B. 1959/1970. "The Imitation of Nature," in *Landscape* 9:1 (Autumn 1959); reprinted in *Landscapes: Selected Writings of J.B. Jackson*, ed. by E.H. Zube. Amherst: University of Massachusetts Press.

Jackson, J.B. 1986a. "The Word Itself," in *Discovering the Vernacular Landscape*. New Haven: Yale University Pres.

Jackson, J.B. 1986b. "A Pair of Ideal Landscapes," in *Discovering the Vernacular Landscape*.

Jackson, J.B. 1995. "In Search of the Proto-Landscape," in *Landscape in America* ed. By G. F. Thompson. Austin: University of Texas Press.

Kunstler, J.H. 1994. *The Geography of Nowhere*. New York: Simon & Schuster.

Lefebvre, H. trans. by D. Nicholson-Smith. 1974/1991. *The Production of Space*. Oxford: Blackwell.

Luccarelli, M. 2010. "The 'Feral' Wilderness: American Studies, Ecoliterature and the Disclosures of American Space," *American Studies in Scandinavia* 42:1 (Spring 2010): 25-46.

Marsh, G.P. 1864/2003). ed. & intro. by D. Lowenthal, *Man and Nature*. Seattle: University of Washington Press.

Marx, L. 1964. *The Machine in the Garden*. New York: Oxford University Press.

Oakes, T.S. & Price, P.L. eds. 2008. *The Cultural Geography Reader*. London: Routledge.

Olwig, K. 2002. *Landscape, Nature and the Body Politic*. Madison, WI: University of Wisconsin Press.

Powell, S. C. 1963. *Puritan Village: The Formation of a New England Town*. Hanover, NH: Wesleyan University Press.

Ruskin, J. 1859. *The True and the Beautiful in Nature, Art, Morals and Religion*. New York: Wiley Halstead.

Tuan, Y-F. 1974/1990. *Topophilia*. New York: Columbia University Press.

II

Anti-Landscapes

5

Modernism and the Creation of Anti-Landscapes: The Valley of Ashes in Fitzgerald's *The Great Gatsby* and Los Angeles in Chandler's *The Big Sleep*

Anna Flügge

This essay claims that a strong strand in American Modernism, mostly practiced by T. S. Eliot and Ezra Pound, sets the pattern for the portrayal of wastelands and anti-landscapes. While these two terms share many similarities, *wasteland* is the broader one, since it also refers to places that have never been inhabited. However, many Modernist texts refer specifically to the city and its lack of qualities that are important for humans, thus creating anti-landscapes, often in a metaphorical sense. This focus expresses the authors' decisive cultural pessimism. Modernism is unthinkable without the wasteland, since it is at the core of its aesthetics. It also gives postmodernism (and posthuman literature) an image to use for its own frequent focus on nuclear and toxic environments. The depictions of anti-landscapes in Fitzgerald's 1925 novel *The Great Gatsby* (with the focus on its valley of ashes) as well as Raymond Chandler's *The Big Sleep* (the oilfields) will be discussed in detail as examples of a group of Modernist texts.
Index: waste land, anti-landscape, *The Waste Land*, city. cultural pessimism, *The Great Gatsby*, *The Big Sleep*, "melancholic Modernism."

The wasteland is a frequent image in Modernist literature. Eliot's 1922 poem of the same name is the best-known and has proffered a trope for the era, as Ronald Berman argues: "The modern movement had after all found its correlative: the great literary and artistic movement of the century's beginnings saw the social world from the urban, dislocated point of view of *The Waste Land*" (82). Apart from Eliot, authors such as William Carlos Williams, who likened the effect of Eliot's poem on American literature to that of an atomic bomb (174), F. Scott Fitzgerald,

Raymond Chandler, and Nathanael West, among others, also depict wastelands in their works. Wastelands, as H. R. Stoneback writes in *Reading Hemingway's* The Sun Also Rises*: Glossary and Commentary*, are "in the general English language sense [. . .] a desolate, barren, or ravaged land, or symbolically, a spiritually or culturally barren or hopeless place" (155). A wasteland expresses the aesthetics of a strand of Modernism. While wastelands often used to have positive connotation, as new and unspoiled spaces, challenging humans to use them for their purposes, this changed in Modernism. Preceded by an international wave of cultural pessimism, which was especially strong in Germany, for example in the writings on civilization by philosopher Friedrich Nietzsche (see Thormählen 136-37), this negative perception of civilization also affected American literature, if somewhat later. The critic Moglen sees extreme and unstoppable capitalism as responsible for the perception that traditional ways of life cannot be preserved, and many Modernist authors mourn this change. The idea of progress is seriously questioned at this time; instead a "realignment of thought" or "a radical alteration takes place to give us the intellectual conventions of Plight, Alienation, and Nihilism; the idea of the modern is bound up with consciousness of disorder, despair, and anarchy" (Bradbury/McFarlane 40-41). The old order did not seem to work; the traditional view of progress had to be abandoned, and a new form that expressed the experience of life as disorder had to be found. As Bradbury and McFarlane write, Modernism responded "to the scenario of our chaos" (27). The destruction of World War I appeared to document the validity of cultural pessimism, and the ravaged battlefields were a literal wasted land.

Wastelands share many traits with anti-landscapes as defined in the introduction to this volume: "An anti-landscape is a man-modified space that once served as infrastructure for collective existence but that has ceased to do so, whether temporarily or long-term. Human beings can inhabit landscapes for generations, even millennia, but they cannot long inhabit anti-landscapes." An anti-landscape was once part of human settlement; the term *wasteland*, on the other hand, is broader, referring to both humanly transformed spaces and to wilderness, such as deserts. Both settings are discussed in Eliot's five-part poem. Instead of seeing empty spaces in a positive light, for example for exploration and new beginnings, the poem describes them negatively. They show the "real" underlying nature of the world as the author perceives it.

The city is a frequent setting in Modernism; its anonymity is deadly to authenticity, and only the masses thrive there, not the Modernist individual. The city, being man-made, becomes an anti-landscape, at least as far as spiritual needs are concerned. The focus on anti-landscapes in Modernist novels such as the ones I am going to discuss is important because it emphasizes decline; it strongly expresses the cultural pessimism felt by Modernist authors. At the same time, in all its desolation, it expresses a new, untraditional aesthetics. As Susan Sontag notes in *On Photography*, people in the Modernist era may see photographs of bleak city scenes and traditional pictures as being equally appealing, and might even prefer the former (78). The modern condition has affected depictions of landscape, which serve as an exteriorization of an inner state of crisis.

Before turning to *The Great Gatsby*, I will discuss Eliot's poem in more detail, focusing on the end of the first part, "The Burial of the Dead," on the third and central part, "The Fire Sermon," and a portion of the fifth part, "What the Thunder Said," all of which focus on the city as an anti-landscape. The poem gave a name and a definition to Modernism's perception of its condition, and the form mirrors the fragmentary character of the world it depicts, as the wasteland contrasts with fragments of the past. To some degree, the past Eliot had in mind was ancient, as he worked with ideas about the nature of myth and ritual that had been developed in early anthropology and synthesized into a framework in Sir James Frazer's *The Golden Bough* and Jessie Weston's *From Ritual to Romance.* The degree of his debt to such sources or their accuracy is important, but not to the argument of the present paper. The point in this context is that Eliot imagined that early mankind had more coherent myths that were a living part of daily life. In contrast, the imagery of death drives his poem, which is dedicated to Eliot's late friend Jean Verdenal, whom he greatly admired and with whom he was likely in love, and who died in World War I. Its epigraph is a passage from the fate of Sybil of Cumae, who had asked the Gods for eternal life, but not eternal youth. Hanging in a cage, boys ask her what she wants, and she answers that she wants to die. This passage sets the tone for work as a whole, which is dominated by scenes of decline and isolation. Life has become mere existence. Civilization is in its last stages, but is as yet unable to die; it is waiting for the apocalypse. The condition is ubiquitous; no salvation offers itself; both city and desert are wastelands.

The end of the first section is a bleak city scene in the present, suffused with a sense of apocalypse:

Unreal city,
Under the brown fog of a winter dawn,
A crowd flowed over London Bridge, so many,
I had not thought death had undone so many.
Sighs, short and infrequent, were exhaled,
And each man fixed his eyes before his feet.
Flowed up the hill and down King William Street,
To where Saint Mary Woolnoth kept the hours
With a dead sound on the final stroke of nine.
There I saw one I knew, and stopped him [. . .]
'You! hypocrite lecteur! – mon semblable, – mon frère!' (60-76)

People go to meaningless jobs at dawn, without looking at each other. The desperate attempt at communication fails, which leads the lyrical I to appeal to the reader as a last resort.

The third part of the poem begins with a description of an anti-landscape, reminiscent of the beginning of the first part: "The river's tent is broken; the last fingers of leaf/Clutch and sink into the wet bank. The wind/Crosses the brown land, unheard. The nymphs are departed (173-175). This passage portrays a barren land; it continues with a description of the absence of things that, although they are garbage, signify people spending time by the river in summer: "The river bears no empty bottles, sandwich papers,/Silk handkerchiefs, cardboard boxes, cigarette ends/Or other testimony of summer nights. The nymphs are departed" (177-179). (The lyrical I thinks of the Fisher King, who in the modern world stands for infertility.) Thus, the landscape is dead, and myth has no place. The poem cites from Andrew Marvell's "To His Coy Mistress," "But at my back from time to time I hear" (196), but the familiar words "Time's winged chariot hurrying near" becomes a new version of Day's lines "'A noise of horns and hunting, which shall bring/'Actaeon to Diana in the spring…" (footnote to line 197), namely "The sound of horns and motors, which shall bring/Sweeney to Mrs. Porter in the spring" (Eliot 197-198). Thus, poetic images of the past have lost their meaning; they become ironic in the contemporary world.

Tiresias, a blind seer (and important contrast to Mme Sosostris of the first part) witnesses a scene of loveless lovemaking between a typist and a clerk in a very desolate apartment. The most intimate has become meaningless here, performed with a stranger at the "violet hour" (215), a time that has mythical connotations for him. Tiresias, who has witnessed tragedies on a grand scale, for example King Oedipus' fall, is the "most important personage in the poem, uniting all the rest. [. . .] What Tiresias *sees*, in fact, is the substance of the poem," Eliot writes in the footnote to line 218, "I Tiresias, though blind, throbbing between two lives,…" Tiresias is unable to understand the "indifference" (242) he witnesses. Since he has this function of an observer, it is important to emphasize the discrepancy between the indifference of the characters involved in the scene observed, and the (apocalyptic) imagery used when talking about Thebes. While his experiences in Thebes were horrible, they still convey meaning and heroism. The indifference he witnesses in the world of *The Waste Land* seems worse and makes him speechless.

The fifth part includes a section on antique cities, here standing for great civilizations, and their downfall, juxtaposed with the modern city (of London). Again, the "violet air" (372) refers back to Tiresias' observation in the "violet hour" and to the decline of great civilizations; "Falling towers" are apocalyptic imagery, giving way to a deserted landscape made uninhabitable by aridity, by the lack of water and thus the impossibility of rebirth, and thus turned into an anti-landscape.

The valley of ashes in *The Great Gatsby* is an American wasteland and anti-landscape. The narrator Nick Carraway describes the valley of ashes at the beginning of the second chapter, and he specifically refers to it as a "waste land" in a detailed description (Fitzgerald 30). Critics have linked this scene to Eliot's poem. For John W. Bicknell, it describes the "Valley of Dry Bones" (557). Lionel Trilling comments on the novel's abstract nature; he writes that the characters are "not 'developed,'" but instead

> are treated, we might say, as if they were ideographs, a method of economy that is reinforced by the ideographic use that is made of the Washington Heights flat, the terrible "valley of ashes" seen from the Long Island Railroad, Gatsby's incoherent parties, and the huge sordid eyes of the oculist's advertising sign. (It is a technique that gives the novel an affinity with *The Waste Land*, between whose author and Fitzgerald there existed a reciprocal admiration.) (252)

This anti-landscape is important in regard to Nick's geographical deliberation. He discloses little about his own past – as though ashamed of his origins like Gatsby (Berman 87) – and he justifies his decision to leave the Midwest and go East as follows:

> I graduated from New Haven in 1915, just a quarter of a century after my father, and a little later I participated in that delayed Teutonic migration known as the Great War. I enjoyed the counter-raid so thoroughly that I came back restless. Instead of being the warm centre of the world, the Middle West now seemed like the ragged edge of the universe – so I decided to go East and learn the bond business. (Fitzgerald 9)

Nick's experiences do not allow him to feel comfortable in the Midwest, which no longer seems innocent and hopeful but naïve; he needs to go back in time, to go East instead of West to find a future and a place that answers to his interior landscape. The East is the middle ground between war-torn Europe and the Midwest. This is important in regard to the description of the valley of ashes at the beginning of chapter two. It is inhabited by Myrtle Wilson, Tom Buchanan's mistress. Her husband George Wilson is pale and lacking in will or substance, the human embodiment of this place. This wasteland is "[a]bout half way between West Egg and New York" (29), thus also constituting a middle ground.

The valley of ashes, "a certain desolate area of land," is described in poetic language that indicates a certain beauty in its desolation:

> This is a valley of ashes – a fantastic farm where ashes grow like wheat into ridges and hills and grotesque gardens; where ashes take the forms of houses and chimneys and rising smoke and, finally, of ash-grey men, who move dimly and already crumbling through the powdery air. Occasionally a line of grey cars crawls along an invisible track, gives out a ghastly creak, and comes to rest, and immediately the ash-grey men swarm up with leaden spades and stir up an impenetrable cloud, which screens their obscure operations from your sight. (29)

Ashes dominate and suffocate the land – there is no indication that ashes might also fertilize and become part of a cycle of renewal. They are the waste of a growing city population that needs heating and electricity, and thus a reminder of the long-term unsustainability of ever-growing cities. The ashes seem to take the form of things and people, a negative form of creation, instead of merely covering the objects underneath. This is the substance of the world as it appears to Nick.

Technology dominates the land; the railroad runs through it, and cars drive through it on their way to New York. Also, it is the area of the forgotten signs of the age, the eyes of Doctor T. J. Eckleburg, the advertisement by an oculist that presides over the desolation, and it is later revealed that Wilson sees the eyes of the advertisement as the eyes of God. Consumer culture has thus become the new religion. The valley of ashes is the "dumping ground" (29) of technology and modern capitalist society, and probably an unavoidable consequence of the adjacent vibrant city. It is a modern, perverse version of the American small town. As Nick observes: "The only building in sight was a small block of yellow brick sitting on the edge of the waste land, a sort of compact Main Street ministering to it, and contiguous to absolutely nothing" (30). This place resembles Nick's interior landscape, even if he shuns it, trying not to look at it when he goes to the city. There are no morals here – Myrtle Wilson is picked up by Tom, with Nick as a witness. Later, when the police come after the fatal car accident, it emerges that the place is nameless: "'What's the name of this place here?' demanded the officer. 'Hasn't got any name'" (146). The lack of a name signifies a lack of identity or meaning.

It is also significant how the words used to describe the valley of ashes – for example the "powdery air" and the "impenetrable cloud" brought about by movement there (29) – evoke Nick's disgust with the "foul dust [that] floated in the wake of his [Gatsby's] dreams" (8; see also Moyer 52). At the same time, however, no pretense is being made in the valley of ashes that the world is anything but a bad place, an impression that the characters in the novel try to ignore. Only apparently peripheral, this wasteland stands at the core of the novel. It must be passed going to and from the City, which Myrtle visits to be with Tom, it is also linked to West Egg where Mr. Wilson goes to kill Gatsby, thinking that it was he who ran down his wife. Almost all the characters are hollow and modernist to the core, except for Gatsby and arguably Nick, who to varying degrees hang on to outmoded romantic and pastoral ideals.

A major theme in the narrative is where the roots of this immorality are, that is, where the American project went wrong. While the opening of the book suggests that an ideal America might still be found in the Midwest, the ending places that ideal forever in the past. Yet Nick, who is not always entirely reliable, is rather vague about the turning point or the corruption of the national dream, or about a definite beginning of the process. It seems to be partially due to the imperfections of the dream itself and partially due to the immorality and carelessness of the rich. This might be called the "unconsciousness of mind without will," a phrase which Berman attributes to Eliot (92). The stasis in the lives of most of the characters is apparent. Tom sees adult life as a decline from the heroics of college football and is concerned with maintaining the status quo. He fears losing power due to the rise of people from the lower classes and non-White races. Daisy and Jordan similarly have no idea of what to do with the future. Indeed, Daisy asks the question directly early in the novel, in words similar to those used in part II of Eliot's *The Waste Land*; she utters a "sentence," as Bicknell observes, "symbolic of her emptiness; she is like Eliot's lady in 'The Waste Land' who cries out, 'What shall we do tomorrow? What shall we ever do?'" (559).

Like the lyrical I in *The Waste Land*, Nick turns away from the extravagance and immorality of what he has seen, but unlike the lyrical I, he returns to his native Midwest instead of looking to the East. For a time, Gatsby's idealism made life in the wasteland bearable, or at least so fascinating that Nick remained. After Gatsby's funeral, however, Nick cannot continue lying to himself, as he phrases it when he breaks up with Jordan. This is not a life he can bear.

The ending of the novel comes back to the thin layer of civilization that covers the underlying wilderness condition, or the America that existed before the arrival of Dutch sailors, whom Nick refers to (187). From the perspective of the early explorers, he sees the houses as "immaterial" (187) making it possible to imagine something commensurate to Gatsby's capacity for wonder. In this vision the Midwest, is transformed into empty and not yet violated space. Moreover, even in the present, as Bicknell points out, it is unlike New York, a place "where the Horatio Alger traditions of self-reliance and enterprise still had some vitality" (560). As Leo Marx notes, the ending shows Nick's realization that he will not find what he is looking for in the East or the West, and his going back is a "mere salute to the memory of a vanished America" (364).

> Nick's repudiation of the East is a belated, ritualistic withdrawal in the direction of "nature." It is ironically set against the fact, which the entire novel makes plain, that the old distinction between East and West has all but disappeared. Nick's final gesture is a mere salute to the vanished memory; and the book ends on a sadly enervated note of romantic irony: "So," Nick says, "we beat on, boats against the current, borne back ceaselessly into the past." (364)

Marx's claim that there is no distinction anymore, no new place to go, is supported by Nick's extensive account of the excitement of the Christmas trips home during his school days. Going back to the past is his only option and at the same time a defeat, contrary to the very idea of America.

The vision that this novel reflects is perhaps even darker than that of *The Waste Land*, in that the past Eliot images is real, substantial, and great. Moglen, who divides texts of Modernism into two groups, places both *The Waste Land* and *The Great Gatsby* in the group of texts he sees as representative of "melancholic Modernism," as opposed to the texts of the other category, "Modernism of mourning":

> There are, then, two modernisms in the United States. They emerged alongside one another, in tension and in dialogue. Together, they constituted one of the most important arguments in twentieth-century American culture. This was, centrally, an argument about the suffering that had accompanied modern capitalism – a struggle between those who imagined that the alienation and injustice of modern life reflected grim and unalterable facts about human nature and those who insisted that these inhumane circumstances had been produced by a destructive social order that could be remade. [...] It was an encounter between two modes of response to social crisis and collective injury: an encounter between melancholia and mourning. (XIV)

Thus, both texts reflect a vision that does not see human agency as capable of changing anything. Moglen carries his argument further, claiming that in melancholic Modernism, it is both the condition of the times and timeless unalterable conditions that are responsible for the darkness of the vision: "the bitter alienation produced by modernization is at once a historically particular crisis and the eternal recurrence of an inescapable existential nightmare" (33). While both texts fall into that category, *The Great Gatsby* seems to go further, not mourning a great past, but thematizing the idea of America as an inspiring illusion rooted in a lost transitory moment.

Raymond Chandler's *The Big Sleep* (1939) can be included in the same group of texts to which *The Waste Land* and *The Great Gatsby* belong. California fiction seems especially fitting for the depiction of anti-landscapes; the state and especially the city of Los Angeles have come to stand for this strand of Modernism. A contributing factor is its geographical setting – the utmost west, after which there is no more land and, thus, no more hope for the dream to come true – as well as the existence of the movie industry, which is so pervasive that the world it creates seems to be imitated by "real" life, instead of vice versa. This artificial world lacks a homogenous and thus meaning-giving tradition; instead, the architecture, for instance, is depicted as a wild mixture of copies of different styles from older societies. This becomes especially clear in novels such as West's *The Day of the Locust*, but also in crime novels such as *The Big Sleep*, which I will discuss in detail as one example of a powerful tradition.

The genre of American detective fiction is primarily written for entertainment, but its genre conventions of darkness, urban blight and moral ambiguity offer a vehicle for social commentary, including the depiction of the city as a metaphorical anti-landscape. It picked up the "Modernist tone" and popularized it. Moreover, the genre's filmic possibilities have brought it to a large audience who have learned to see moral ambiguity convincingly portrayed by lighting accompanied by certain forms of music. Likewise, urban blight makes for good images on screen, beginning in the black and white "noir" films. Chandler's Philip Marlowe encounters a corrupt and hostile city, where he often is asked to work for the morally corrupt rich clients, including, in *The Big Sleep*, two careless daughters, Vivian and Carmen. In addition, there are those who can hardly make ends meet, who are driven to a life of crime in order to survive. Crimes in the novel range from blackmail, gambling, prostitution and distribution of pornography to murder. In all, six people are killed. Homosexuality is depicted as very negative, as yet another sign of disorder. The pervasive police corruption – for instance, the police ignore Geiger's pornography sideline even though his bookstore is located prominently (30) – suggests that any hope for improvement is misguided. Marlowe is a modern Tiresias, often more an observer than a participant. He is educated but poor, and highly critical of the rich and the corruption they stand for. He is thus almost classless. Marlowe's view of Los Angeles as an anti-landscape is inescapably linked to the corruption of the rich, including their methods of making money. The novel shows an interesting perspective on the American past: the present is a consequence of the social system and it seems to have existed from the beginning of history. There is no moment of purity like the one of seeing the new world for the first time described at the end of *The Great Gatsby*, and the system is becoming more and more powerful, creating an anti-landscape when it is growing, as I will show later.

Los Angeles civilization is insubstantial, as shown symbolically in the description of the houses; as Marlowe states when he tries to force open a front door: "About the only part of a California house you can't put your foot through is the front door" (34). The city has also become an impersonal force, a logical consequence of exploitation. It is constantly growing, accelerating the inherent decline. As Marlowe says to his antagonist Eddie Mars: "This is a big town now, Eddie. Some very tough people have checked in here lately. The penalty of growth" (73). Perhaps more than Gatsby, the detective is a romantic hero who is out of place in a modern world that does not have any use for his values. (Gatsby did make his money in bootlegging and selling stolen bonds, after all.) While attracted to his client's beautiful older daughter, Vivian, Marlowe has his code that determines his actions. His code puts the client first, and it is often directly opposed to the police's investigative methods. The code helps Marlowe find a way through the maze of modern life, but it also isolates him.

The anti-landscape in this novel is also literal. Below the Sternwood mansion are the mostly defunct oil fields that generated the family's wealth and that, as Marlowe suspects, the family no longer wishes to see:

> Beyond the fence the hill sloped for several miles. On this lower level faint and far off I could just barely see some of the old wooden derricks of the oilfield from which the Sternwoods had made their money. Most of the field was public park now, cleaned up and donated to the city by General Sternwood. But a little of it was still producing in groups of wells pumping five or six barrels a day. The Sternwoods, having moved up the hill, could no longer smell the stale sump water or the oil, but they could still look out of their front windows and see what had made them rich. If they wanted to. I didn't suppose they would want to. (21)

This passage reiterates the point that exploitation has been part of the dream since the beginning. The rich can make a good life if they are willing to damage their own properties. As David Wyatt writes, "the activity of extraction defaces a Southern California Eden even as it finances it" (36). Marlowe encourages the reader to share his sentiment when he later declares, "To hell with the rich. They made me sick" (64).

Their family having exhausted the land – Vivian speaks of the recklessness of the Sternwoods in all areas of life (59) – the daughters are now engaged in gambling, drinking alcohol, and a dissolute life. Marlowe is disgusted with a woman who, in spite of the death of two men with whom she was having relationships, only thinks of herself. He has to witness the death of one of them, who shows honor and protects her. The place of the murder, the Fulwider Building, mirrors the horror of the crime and is the architectural opposite of the Sternwood mansion, representing the dark and bleak side of Los Angeles. The building is a world of have nots whose lives are hopeless:

> A single drop light burned far back, beyond an open, once gilt elevator. There was a tarnished and well-missed spittoon on a gnawed rubber mat. [. . .] Numbers with names and numbers without names. Plenty of vacancies or plenty of tenants who wished to remain anonymous. Painless dentists, shyster detective agencies, small sick businesses that had crawled there to die, mail order schools that would teach you how to become a railroad clerk or a radio technician or a screen writer – if the postal inspectors didn't catch up with them first. A nasty building. A building in which the smell of stale cigar butts would be the cleanest odor. (170)

Marlowe later kills the murderer but it only makes him feel better for a short time (203). He has to let Eddie Mars go free – even though Mars is the mastermind behind many of the crimes – in order to protect his client, General Sternwood. Carmen, the lonely and neglected younger daughter who has murdered one man and tried to kill Marlowe as well, is merely sent to an institution. His decision not to hurt his dying client, which he would if he told him what he has found out, means that the case remains officially unsolved, and Marlowe has to live with his newfound knowledge. Carmen's sister Vivian, who knew about it all but lies as long as possible and then tries to bribe Marlowe, remains in the mansion, in her lavish quarters, "A screen star's boudoir, a place of charm and seduction, artificial as a wooden leg" (221), as Marlowe describes it.

At the end Marlowe drives "off down the hill" (230) to downtown Los Angeles, symbolically leaving the elevated realm of the rich and returning to the other side. He realizes that he used to be an observer and thus in a way pure, keeping the city and the women about to corrupt him away, but has become "part of the nastiness now" (230). He then goes to a bar on the way, drinking and missing the villain's beautiful and devoted wife, remarking that he "never saw her again" (231). This ending shows a desperate acceptance and complicity, as well as the impossibility and futility in trying to get out, since the frontier is closed. The detective, the urban cowboy, is mired in the anti-landscape while trying to escape into a form of fragile inner exile. And he is obsessed with the city, with solving its mysteries.

The Big Sleep is a further example of melancholic Modernism. Even though action is taken by killing one criminal, it does not change anything. Unlike the texts mentioned earlier, Marlowe does not try to flee, but rather accepts the reality and inevitability of his situation, becoming part of the city and its darkness. Thus, Chandler's vision is still darker than that in *The Great Gatsby*. The city as an anti-landscape swallows the protagonist, and the fact that *The Big Sleep* is only the first of five novels that feature Marlowe as the protagonist suggests that Marlowe is right about there being no escape.

The strand of Modernist literature discussed above focuses on the anti-landscape. While the anti-landscape is often of a rather metaphorical nature (being bereft of meaning makes it uninhabitable, or spiritually wasted), the texts' cultural pessimism and the form given to that pessimism have provided authors with a poetic vision which during and after the Second World War was expanded and developed into the vision of more thoroughly devastated and more extensive wastelands.

Bibliography

Berman, Ronald. 2002. "*The Great Gatsby* and the Twenties." *The Cambridge Companion to F. Scott Fitzgerald*. Ed. Ruth Prigozy. Cambridge, UK: Cambridge University Press, 79- 94.

Bicknell, John W. "The Waste Land of F. Scott Fitzgerald." *The Virginia Quarterly Review* 30 (1954): 556-572.

Bradbury, Malcolm and James McFarlane, eds. 1991. Introduction. *Modernism: A Guide to European Literature 1890-1930*. 1976. London: Penguin. 19-55.

Chandler, Raymond. 1992. *The Big Sleep*. 1939. New York: Vintage.

Eliot, T. S. 1974. *Collected Poems 1909-1962*. London: Faber.

Fitzgerald, F. Scott. 1994. *The Great Gatsby*. 1926. London: Penguin.

Marx, Leo. 2000. *The Machine in the Garden: Technology and the Pastoral Ideal in America*. 1964. Oxford, UK: Oxford University Press.

Moglen, Seth. 2007. *Mourning Modernity: Literary Modernism and the Injuries of American Capitalism*. Stanford: Stanford University Press.

Moyer, Kermit W. 1972. "*The Great Gatsby*: Fitzgerald's Meditation on American History." *Fitzgerald-Hemingway Annual*, 43-57.

Nye, David. 2010. *When the Lights Went Out: A History of Blackouts in America*. Cambridge, MA: MIT Press.

Sontag, Susan. 2001. *On Photography*. 1977. London: Picador.

Stoneback, H.R. 2007. *Reading Hemingway's* The Sun Also Rises: *Glossary and Commentary*. Kent, Ohio: Kent State University Press.

Thormählen, Marianne. 1978. *The Waste Land: A Fragmentary Wholeness*. Lund, Sweden: CWK Gleerup.

Trilling, Lionel. 1961. „F. Scott Fitzgerald." *The Liberal Imagination: Essays on Literature and Society*. By Lionel Trilling. 1950. London: Mercury. 243-254.

Williams, William Carlos. 1951. *The Autobiography of William Carlos Williams*. New York: Random House.

Wyatt, David. 2010. "LA Fiction through Mid-Century." *The Cambridge Companion to the Literature of Los Angeles*. Ed. Kevin R. McNamara. Cambridge, UK: Cambridge University Press. 35-47.

6

A Landscape of Fear: Cormac McCarthy's *The Road*

Øyunn Hestetun

Sometimes faith might just be a case of not havin nothin else left. (McCarthy, *The Sunset Limited* 118)

Cormac McCarthy's *The Road* (2006) features a dystopian vision of a post-cataclysmic world in which the physical environment has been turned into an anti-landscape and humans have turned to savagery. The story focuses on a father and son who are heading south in search for food and shelter, wandering through a wasteland that takes the contours of an inverted pastoral and Yi-fu Tuan's "landscape of fear." A spark of hope is retained in the motif of the carriers of fire, suggesting that the novel could be defined as a critical dystopia, while it also alludes to other literary genres and modes, such as the road narrative, the parable, and the post-pastoral. This essay argues that the novel conveys an ecopoetic or ecodystopic sensibility that may sharpen our awareness of dependence on the world we inhabit, at the same time that it presents a meditation on metaphysical concerns.

Key names and concepts: Mikhail Bakhtin, Frederick Buell, Lawrence Buell, Cormac McCarthy, Yi-fu Tuan, landscape, anti-landscape, critical dystopia, parable, road narrative, post-pastoral, environmental concerns, metaphysical speculation.

1. Introduction

The image of an anti-landscape – "Barren, silent, godless" – is conjured forth already in the first pages of Cormac McCarthy's *The Road* (2006). The novel opens with a description of a nameless character waking up in the early morning from a nightmarish dream, and while his child is still sleeping, he leaves the makeshift shelter where they spent the night to survey the area. Through his binoculars he observes "soft ash blowing in loose swirls over the blacktop" and "the dead trees," and for a while he sits there, "watching the ashen daylight congeal over the land" (McCarthy 2006: 2-3).[1] The landscape thus introduced and further pondered in the pages that follow does not match the common conception of a sustainable human habitat. Rather, *The Road* features a dystopian vision of a future world in which the physical environment has been transformed into a wasteland and civilization has regressed into barbarism. In view of John B. Jackson's definition of landscape as "a composition of man-made or man-modified spaces to serve as infrastructure or background for our collective existence" (1984: 8), it is obvious that the landscape in McCarthy's *The Road* no longer fulfills such a function, and that it could more appropriately be called an anti-landscape. Yet the novel's anti-landscape is perhaps less the representation of an actual place or landscape than an imaginary setting of a tale verging on allegory. The narrative details the journey through the devastated landscape of an unnamed father and son who are headed towards the warmer climes of the south, seeking refuge from a hostile environment of freezing cold, lack of food for sustenance, and humans turned into cannibals. While the tale might be considered an allegorical meditation on existential questions of good and evil, hope and despair, its dystopian vision of a world stricken by disaster of cataclysmic proportions also feeds into ecological concerns about the direction in which our civilization may be headed.[2]

Disaster, horror, violence, and human depravity are common ingredients in McCarthy's fictional universe. Like so many of his earlier novels, *The Road* could be said to draw on the tradition of the Southern Gothic. But with this book the author stakes out a new direction in his writings.[3] Most obviously, rather than looking back at the region's recent past or the more distant history of the Southeast or the Southwest, this novel

1. Subsequent references to McCarthy's *The Road* will include only page numbers.
2. While critics debate to what extent it is possible to find an ecocritically informed critique of contemporary society in the novel, Sara L. Spurgeon notes that "ecocritical concerns," traceable in McCarthy's earlier works, are "fully realized in *The Road*" (2011: 12), and George Monbiot, one of Britain's leading "green commentators," deems it to be "the most important environmental book ever written" (2007: n.p.).
3. See e.g. Rune Graulund or Steven Frye (2009: 166-67).

projects an indeterminate future and a non-specified place.[4] The narrative is rendered in a minimalist prose that contrasts with the expansive and verbose style of the bulk of the author's earlier works, while retaining the distinctly poetic quality, and a language that resonates with biblical allusions. Most importantly, the McCarthian grotesque has, as it were, been extended to the descriptions of the natural world. I will explore how the novel outlines an anti-landscape of fear, while drawing on elements borrowed from different generic modes, ranging from the road novel, the parable or the allegory, the inverted pastoral, and the dystopia.

2. An Inflection of the Road Narrative

The novel's title brings to mind Mikhail Bakhtin's chronotope of the road as described in his "Forms of Time and of the Chronotope in the Novel."[5] Central to the Greek adventure romance, the road chronotope typically features the picaresque hero's travel through a varied landscape, depicting "random encounters" between people representing different walks of life (Bakhtin 1990: 243). While the basic structural pattern of the road chronotope is retained, the defining motifs are radically transformed, as the picaresque has been exchanged with the grotesque, and the familiar territory has become a landscape of fear. McCarthy's title likewise evokes the more recent counterpart in the tradition of the American road narrative, and the novel could be considered as yet another inflection of the genre, albeit staking out a new and different course.[6] Consider Sal Paradise in Jack Kerouac's *On the*

4. Critics that have tried to determine the setting based on references in the novel seem to agree with Frye, who suggests that it is set somewhere in "southeastern United States" (2009: 168). See also Wesley G. Morgan's detailed argument that, "McCarthy is fictionally returning once again to his own roots in Knoxville and the southeast" (2008: 46).
5. Adopting the term "chronotope" to designate the specific manner in which time and space intersect in narrative, Bakhtin suggests that chronotopes may serve as a basis for "generic distinctions" (1990: 85), as they function as "the organizing centers for the fundamental narrative events of the novel" (1990: 250).
6. Commenting on Jim Crace's *The Pesthouse* and McCarthy's *The Road*, Susan Balée observes: "Both writers turn the American road novel on its head by putting survivors – without cars – on the enduring pavement" (2007: 518). See also Susan Kollin.

Road, for instance, who is full of hopeful expectations as he is heading west with his mate behind the wheel: "I could hear a new call and see a new horizon, … Somewhere along the line I knew there'd be girls, visions, everything; somewhere along the line the pearl would be handed to me" (Kerouac 1976: 8). No trace of the optimism, desire for adventure, and hope for fulfillment expressed by Kerouac's protagonist can be found in McCarthy's novel. In contrast to the prototypical American road novel, this journey of father and son is undertaken not on wheels, but on foot; the ultimate goal is not adventure, but survival; and chance encounters on the road induce not excitement, but fear. Most notably, the picaresque hero of earlier versions of the road narrative – or the hipster of its more recent American counterpart – has been transformed into a nurturing, caring father. Even the road itself takes on new significance, as it no longer signals freedom and endless possibility, but rather a condition of life in a disrupted world.

Although father and son steer their course through the desolate landscape by studying a tattered road map, they stay off the road itself as much as possible, so as not to become easy targets of roving bands of marauders and killers. At one point the father reasons: "He thought the road would be so bad that no one would be on it but he was wrong" (107). On the last pages of the novel, after the father's death, the boy encounters a man who warns him: "If you stay you need to keep out of the road," and he adds: "I don't know how you made it this far" (303). This man, who travels with his own family, knows the dangers of the road, and at the end of the novel the boy joins them, but only after he is assured that they are good people who "dont eat people" (304).

In other words, the celebration of adventures in the conventional road narrative has been turned into a harrowing tale of escape, where danger lurks behind each new bend in the road. At first glance the southward journey of father and son could be said to replicate the westward journey of the pioneers of the prototypical American frontier narrative, but with a diverting twist, staking out a course not west but south. What is more, the ideal of national progress is inverted, as the forward march of civilization has been turned into regress. Nor do we find a parallel to Huck Finn's "lighting out for the territory," in an act of fleeing domesticating civilization, and seeking freedom and adventure. For father and son in McCormac's story, the journey is a quest for survival, and an escape from death by starvation and the freezing cold of winter.

3. A Landscape of Fear

McCarthy's narrative of father and son whose existence has become a continuous struggle for survival conveys a strong sense of the fear or terror that humans may experience in their encounter with the natural world.[7] In the description of the "wasted country" (4) covered by gray ashes that the father contemplates in the opening pages, it is possible to recognize the contours of the kind of anti-landscape that Yi-fu Tuan outlines in his *Landscapes of Fear*. Tuan comments on the feeling of fear and terror that may be induced by an encounter with the natural world. "Fear is in the mind," he writes, but "has its origins in external circumstances that are truly threatening." Hence, he explains, the expression "'Landscapes of fear' refers both to psychological states and to tangible environments" (Tuan 1980: 6). He discerns two basic sensations, of which one relates to "personalized evil," taking the shape of "hostile forces," and the other is described as "fear of the imminent collapse of his [sic] world and the approach of death – that final surrender of integrity to chaos" (Tuan 1980: 7). In McCarthy's novel fear is a constant sensation and a preoccupation of the father and son as they proceed on their way through the ruined landscape. They have to be on constant guard against Tuan's "personalized evil," which in the novel is given human shape. These humans turned monsters could be regarded as degenerates, or evidence of how the human species may revert to basic instincts in a struggle for survival in a post-cataclysmic wasteland. They could also be considered players in an allegorical battle between the forces of good and evil in a seemingly godless universe. Thus the main plot line of the novel could be considered to be based on a quasi-scientific perspective, complemented by a religious-mythological perspective.

This religious-mythological perspective is further developed at different levels of the narrative, above all in dream sequences, in the imagery of narratorial asides, and in conversations. In the dream that the father wakes up from in the opening scene of the novel, for instance, father

7. Balée comments on what she calls "the claustrophobic atmosphere of fear and anxiety" (2007: 520), and finds that the novel's "mood is Gothic, but distinctly American." She draws the lines back to Poe's "The Fall of the House of Usher," except that in McCarthy's tale, "the whole country is his ruined edifice"; and while Poe "plays on the taboo of incest," McCarthy draws on "the human fear of cannibalism" (2007: 518).

and son take on the role of "pilgrims in a fable swallowed up and lost among the inward parts of some granitic beast" (1). While in the womb of the beast, which in itself represents an instance of Tuan's "hostile forces," they also encounter another beast: "a creature that raised its dropping mouth from the rimstone pool," and "swung its head from side to side and then gave out a low moan and turned and lurched away and loped soundlessly into the dark" (2). One might speculate whether these beasts are inspired by the biblical Book of Revelation or whether they hark back to mythological creatures of chaos. Regardless, such beasts represent personifications of the powers of chaos: chaos is indeed let loose on the fictional world constructed in the novel and is not confined to the world of dreams. The dream recounted in the opening paragraphs foreshadows the chaos, horror, and adversities that father and son will encounter on their journey.

The sensation of fear is evoked also by the external landscape, which has turned into an anti-landscape. The terrain that father and son wander through is fundamentally dead, devoid of regenerative force. The land is no longer a sustainable source and foundation of vegetative, animal, or human life. The familiar territory has been totally de-familiarized. Driven from the ruins of what was once home, father and son are headed south in search of food and a warmer climate. The fields have been turned into stubble of "dead grass" (111), and the forests consist of "trees dead and black" (99). One night father and son are awakened by the sound of a "long dry crack of shearing limbs" (101), as dead trees – weighed down by the falling snow – come crashing down around them. Crossing a river, they look down on "the slow dead water passing underneath" (169). Near an abandoned farm the father observes the remains of earlier days: "Dead ivy along a stone wall and a mailbox and fence along the road and the dead trees beyond. Cold and silent. Shrouded in the carbon fog" (124). When they finally reach the coast, it is only to find that the ocean has not been spared: "At the tide line a woven mat of weeds and the ribs of fishes in their millions stretching along the shore as far as eye could see like an isocline of death. One vast salt sepulcher. Senseless. Senseless" (237). Death and destruction are everywhere, and everything is covered in ashes. It is as though the natural cycle of life and death has come to an end.

As the landscape is no longer fit to serve as a human habitat, human civilization has also regressed. The fear evoked by the barren landscape is paralleled by the total collapse of community and society, and extends to the degenerative behavior of human beings who leave death in their wake, as humans have turned to slavery and cannibalism. Inspecting "the remains of an orchard," with trees "gnarled and black and the fallen limbs thick on the ground," the father detects in the "stubble grass" the bloodstains of the slain having been "field-dressed and hauled away" (94). While the people they fear have stooped to basic instinctual behavior in their struggle for survival, the father and son seek to preserve their human dignity and hold on to values of right and wrong. The father tries to spare his son the most atrocious displays of horror – based on the principle that "the things you put into your head are there forever" (11) – but he also realizes that his own fear rubs off onto the child. Accordingly, he sets himself the task of teaching his child not only strategies of survival and means of defense, including suicide, but also strategies for how to handle his fear. In this post-cataclysmic world fear is a constant companion, and the threats of the ultimate end of the world, chaos, and death are real and tangible. The conversations between father and son circle around the day-to-day struggle for survival, how to find food for sustenance, and how to steer their way through the devastated landscape without taking too many risks, so as to avoid exposing themselves to dangerous forces in the shape of humans. They keep fear at bay by repeating to themselves and to each other that they are okay, that they are and will remain good, and that they will not succumb to the barbarous behavior of the bands of marauders that they encounter on their way.

4. An Ecodystopian Parable, or a Post-Pastoral

With its depiction of an anti-landscape – or Tuan's landscape of fear – the novel lends itself to be aligned with a literary trend that gained momentum in the wake of Rachel Carson's *Silent Spring* from 1962. The tremendous impact of Carson's book is generally recognized in ecocriticism and environmental theory, and the book is even considered to have inspired new directions in fictional writing. Variously referred to as "ecodystopian fiction" (Frederick Buell 2003: e.g. 247), or as "ecocatastrophe novels" or "toxic dystopian fiction" (Lawrence Buell 1998: 649, 649 note 31), a distinctive feature of this genre or mode of fictional writing is a focus on what Frederick Buell refers to as "ecological poisoning and breakdown"

(2003: 247), or what Lawrence Buell calls "toxic discourse."[8] The fairy-tale-like opening of Carson's first chapter from *Silent Spring*, entitled "A Fable for Tomorrow," clearly evokes a pastoral landscape: "There was once a town in the heart of America where all life seemed to live in harmony with its surroundings." The idyll is soon invaded and disrupted, however: "Then a strange blight crept over the area and everything began to change. Some evil spell had settled on the community" (Carson 1963: 3). The rural community formerly ruled by peace and harmony is transformed into a threatening anti-landscape, haunted by "a shadow of death": the vegetation is "browned and withered"; birds, fish, animals, pollinating insects die; humans die from mysterious diseases; and silence rules (Carson 1963: 3-4).

McCarthy's *The Road* could be considered a dark sequel to Carson's "Fable." In a fashion similar to her short preamble, McCarthy's novel imagines – in a fictional form verging on parable or allegory – a future world transformed by global destruction.[9] It is a world of "Nights dark beyond darkness and the days more gray each one than what had gone before. Like the onset of some cold glaucoma dimming away the world" (1). Echoes of Carson's pastoral idyll transformed into a blighted, silenced world can be found, for instance, in the father's reminiscences of a natural environment teeming with birds and other living creatures, set up against the present state of affairs: "Once in those early years he'd wakened in a barren wood and lay listening to flocks of migratory birds overhead in that bitter dark. Their half muted crankings miles above where they circled the earth as senselessly as insects trooping the rim of a bowl. He

8. Originally published under the title of "Toxic Discourse" in 1998, a revised version of the essay appears as the first chapter of *Writing for an Endangered World: Literature, Culture, and the Environment in the U.S. and Beyond* (2001). Noting that toxicity represents "an emergent literary focus" (1998: 665), Lawrence Buell finds that, "Contemporary toxic discourse effectively starts with Rachel Carson's *Silent Spring*" (1998: 645).
9. Several critics point to the allegorical quality of the novel. Spurgeon, for instance, notes its "strongly allegorical atmosphere" (2011: 16). Balée insists, in reference to the novel's "primitive moral power" reminiscent of biblical stories, that "*The Road* is not a satire, it's an allegory" (2007: 520). Frye refers to several reviews that noted its biblical allusions and referred to it as a "parable," and suggests: "In the end *The Road* is set within an imagined but wholly plausible human history, emerging as a parable of human striving that takes form in the journey of a father and son" (2009: 166-67, 179).

wished them godspeed till they were gone. He never heard them again" (54-55). In a conversation between father and son, the father explains that crows now exist only in books, and the boy, who has never even seen a bird, imagines that, "If you were a crow you could fly up high enough to see the sun" (167-77). If the birds are no longer heard or seen in the sky, the fish have likewise vanished from the streams. In an early passage the father ruminates over what has become of the world he knew, looking into the waters of a stream: "He stood on a stone bridge where the waters slurried into a pool and turned slowly in a gray foam. Where once he'd watched trout swaying in the current, tracking their perfect shadows on the stones beneath" (30).

With her fable of a small town "that awakens to a birdless, budless spring," Carson introduces the elements that Lawrence Buell finds to be the "key discursive motifs" or topoi of "toxic discourse" (1998: 645). Buell singles out four topoi distinctive of toxic discourse, which can all be traced in McCarthy's novel. First, "toxic discourse" tends to present "images of community disruption" which "revive a longstanding mythography of betrayed Edens" (Lawrence Buell 1998: 645, 647). As I will illustrate below, an inverted pastoral ideal represents a core motif in *The Road*: the lush, green, fertile landscape of bygone times belongs to the realm of memories and dreams, and has been exchanged for "an ashen scabland" (14), in which "everything [is] dead to the root along the barren bottomlands" (20). In the blighted landscape of the novel we also find a variation on the second topos, that of "images of a world without refuge from toxic penetration" (Lawrence Buell 1998: 648). The father and son harbor little hope of escaping their plight and reaching a place unaffected by the catastrophe; what keeps them going is the realization that they will not survive the cold of the oncoming winter. Even an adaptation of Buell's third distinctive feature – "the moral passion of a battle between David and Goliath" (Lawrence Buell 1998: 651) – can be traced, although it is the victims, above all represented in the father and son, who take center stage. Since the forces that caused the disaster are not specified, the antagonist Goliath could be said to be featured in the threat of violence from marauding bands. These degenerates are also part of the gothic mode, yet another typical feature of "toxic discourse," as are descriptions of distorted bodies, bearing signs of having been caught in the flames that followed in the wake of that first "long shear of light" (54). Other gothic elements include evidence of human cannibalism, such as the sight of "a charred human infant headless and gutted and blackening on the spit" that father and son come upon among the barren trees (212).

While DDT – the "white granular powder" descending on the landscape (Carson 1963: 4) – is singled out as the source of toxification in *Silent Spring*, the cause of the cataclysmic transformation of the natural world is not identified in McCarthy's *The Road*. We only find references to a catastrophe, ostensibly of cosmic proportions, as in the flashback where the father recalls the moment of disaster: "The clocks stopped at 1:17. A long shear of light and then a series of low concussions" (54). Carson's fable also points an accusing finger at those who are to blame: "The people had done it themselves" (1963: 4). The catastrophe in McCarthy's novel, on the other hand, may or may not have been brought on by human intervention, which is to say that a reading of *The Road* as an ecological warning, or a cautionary tale drawing on the topoi of "toxic discourse," would rest on the assumption that human agency is involved in some way or another. Tim Blackmore is in good company when he suggests that the novel describes a world in the grip of a "nuclear winter" (2009: 22), which is "human-created," and which brings into focus "the utterly poisoned quality of the whole ecology" (2009: 24). Accordingly, he argues, the novel can be read as "a kind of warning that might remind us where our technological trajectory will land us" (2009: 18). Others are less rash in their conclusions, however, as Steven Frye who opens up for the possibility that the disaster may have been caused by a "meteor or asteroid," and thereby "the outcome of natural evil" rather than "[a] nuclear holocaust," which would be "the result of human evil" (2009: 169; see also Phillips 2011: 177). Rather than a warning, insisting that civilization is headed for catastrophe unless humans mend their ways, the novel could thus be considered an allegorical meditation on human nature, on the battle between the forces of good and evil, but also on human compassion, love between father and son, and the struggle to retain hope and religious belief in the face of evil and destruction. Interestingly, surveying what has been published in the decades after Carson's book, Frederick Buell finds that "environmental crisis discourse has left apocalyptic ecologism and doomsterism behind and moved *beyond* apocalypse into a variety of new conceptual spaces and rhetorics" (2003: 35). His metaphor "dwelling within crisis" (2003: 205) brings into focus how the more recent trend indicates that the sense of crisis has been internalized, and how the new focus of attention is an exploration of strategies of survival in a contaminated environment.

If the novel can be seen as an inflection on the road narrative or an extended parable, it can also be seen as an inverted pastoral, engaging in Lawrence Buell's "rhetoric of pastoral betrayal" (1998: 649), which was one of the defining topoi of "toxic discourse." Looking into how McCarthy's writings relate to the pastoral tradition, Georg Guillemin, in *The Pastoral Vision of Cormac McCarthy* – which appeared in 2004 and therefore does not consider *The Road* – argues that McCarthy's pastoral vision is not informed by nostalgic desire for a return to a pristine bucolic state (2004: 6), but by what he terms "ecopastoralism," which "bespeaks a worldview whose egalitarian underpinnings logically transcend the preference of the pastoral imagination for natural landscape settings" (2004: 15).[10] In his study *Pastoral* (1999), Terry Gifford notes that McCarthy, in his most recent writings to date, engages in "anti-pastoral constructions of the West," which he links to a long tradition of critical and satirical scrutiny of the pastoral view of nature (1999: 120).[11] However, what we see in *The Road* is perhaps as much informed by yet another mode, the post-pastoral, which Gifford refers to as a "mature environmental aesthetics," which "has gone beyond the closed circuit of pastoral and anti-pastoral to achieve a vision of an integrated natural world that includes the human" (1999: 148).[12]

In this context, one might consider Bakhtin's chronotope of the idyll, which may reappear in fragmented and mutated form in later genres. As "an underlying image" in the development of the novel it has not been sufficiently recognized, he claims, especially in cases that feature what he calls "the destruction of the idyll" (Bakhtin 1990: 228, 229; see also

10. John M. Grammer argues that even in his earlier works, "McCarthy participates in the South's second-oldest intellectual tradition, that of the anti-pastoral" (1992: 20-21). Grammer sees this as a result of the author's desire "to question the old southern dream of escape from history," which is motivated by a fascination with what he deems to be a "characteristically southern delusion," namely, "the pastoral will to create a timeless order" (1992: 21).
11. Accordingly, Gifford notes, McCarthy could be considered, together with writers such as Edward Abbey, as part of a tradition distinguished by "a strain of satires of the pastoral," intended as a critique or a corrective to the pastoral convention (1999: 120, 3).
12. Gifford discusses the defining criteria of the post-pastoral in his chapter 6. No attempt will be made here of checking these against McCarthy's novel.

235). Notably, while Bakhtin delineates adaptations to the modern world (urban life, technology), McCarthy's novel suggests that the adaptation can be made to fit a post-cataclysmic world. Interestingly, Bakhtin's idyll may contain what he calls "philosophical sublimation," in which "the deep *humanity* of idyllic man himself and the humanity of his human relationships are foregrounded, as is the *wholeness* of idyllic life, its organic link with nature" (Bakhtin 1990: 233). McCarthy's "philosophical sublimation" can be traced especially in passages focalized on the father, containing dream images, and flashbacks of memories from the past, where nature is rendered in a pastoral mode verging on the poetic. One such sequence ends with the words: "This was the perfect day of his childhood. This the day to shape the days upon" (12). Passages like these are juxtaposed with presentations of an inverted idyll, or the "destruction of the idyll," as in the following ruminations: "Dark of the invisible moon. The nights now only slightly less black. By day the banished sun circles the earth like a grieving mother with a lamp" (32). The "destruction of the idyll" is brought out in passages focalized on the father, revealing his religious and existential despair; in dialogues between father and son, dwelling on their constant fear; and in sections with zero focalization in which the narrator places the present situation into a larger perspective.

5. A Critical Utopia: "carrying the fire"

In McCarthy's *The Sunset Limited*, which appeared the same year as *The Road*, the born-again ex-convict named Black tries to convince the nihilistic suicidal professor White that, "Sometimes faith might just be a case of not havin nothin else left" (McCarthy 2010: 118). In spite of the obvious differences between these two works, there is a pronounced thematic link between them as both are fundamentally allegorical in form, and contain extended conversations centering on the meaning of life, suffering, religion, hope and despair. Edwin T. Arnold argues that McCarthy's writings always exhibit what he refers to as "a profound belief in the need for moral order, a conviction that is essentially religious," and claims that, "There is … always the possibility of grace and redemption even in the darkest of his tales" (1992: 31). Hope and belief – or the desire to believe – are ingrained also in *The Road*.

Flashbacks reveal that the boy's mother, faced with the prospect of a life dominated by fear and no hope for a better future, has chosen suicide (57-60). The father has chosen differently, and the thought passes through his mind that his wife was probably right in deeming that "the boy was all that stood between him and death" (29). Justifying his murder of a man to save the boy's life, the father explains: "My job is to take care of you. I was appointed to do that by God. I will kill anyone who touches you. Do you understand?" (80). Each day is a struggle for survival, and prepared for the worst with two bullets in the revolver to end their own lives rather than falling victim to the forces of evil, the dying father still retains a spark of hope, which he seeks to keep alive in his son. The motif of hope is present from the start, introduced right after the dream sequence featuring the beasts of chaos and the father surveying the lay of the land. The narrator's comment "He knew only that the child was his warrant," is followed by the words: "He said: If he is not the word of God God never spoke" (3). The biblical "word of God" suggests life and the messianic hope of human redemption through God's grace,[13] and it is as if the father invests his son with messianic expectations. The same idea is repeated in a later statement, which apparently renders the father's thoughts: "Golden chalice, good to house a god" (78), and in the question that the father puts to the man who calls himself Ely: "What if I said that he's a god?" (183). Implicit in these statements is the notion that the child, born after the disaster and with no knowledge of the pre-disaster world other than what his father has told him, is pure, innocent, and untainted by whatever may have been the cause of the present state of chaos and disorder. The question of God's existence resurfaces in the conversation with the old man who calls himself Ely (171-85), who, impersonating a latter-day prophet of nihilism, claims that, "There is no God and we are his prophets" (181), and declares: "Where men cant live gods fare no better" (183). The possibility of human redemption is above all contained in the motif of "carrying the fire," which repeatedly appears in the dialogues between father and son. Leaving biblical connotations aside, the motif of "carrying the fire" brings into focus how

---

13. See John 1.1: "In the beginning was the Word, and the Word was with God, and the Word was God" (NRSV). Both the translation and the interpretation of this verse are debated, and one can only speculate what McCarthy may have intended by his biblical allusion.

father and son represent the "good guys" who do not kill their fellow beings, as opposed to the "bad guys" who do. The fire could be said to symbolize the hope of life, survival, and redemption, and it is significant that the task of "carrying" on this symbol is entrusted to an innocent child.

In this hope resides the key to how the novel could be classified as a critical dystopia, as Tom Moylan defines it in his *Scraps of the Untainted Sky* (2000). Commenting on a turn in utopian and dystopian writing at the end of the twentieth century, he notes that "the dystopian strategy to map, warn and hope" is combined with "a utopian horizon that appears in the text – or at least shimmers just beyond its pages" (Moylan 2000: 196). As a critical dystopia, McCarthy's novel aligns itself with works such as Octavia Butler's the *Parable of the Sower* (1993) and Margaret Atwood's *The Year of the Flood* (2009), which both paint anti-landscapes of a dystopian future, informed by ecological awareness and containing an element of critique whilst at the same time allowing for hope of change and reversal of the forces that caused ecological devastation. Interestingly, the father and son in McCarthy's *The Road* live by a moral code that largely corresponds to the one that Lauren spells out in Butler's *Parable of the Sower*: "… we don't kill unless someone threatens us, … We don't hunt people, We don't eat human flesh. We fight together against enemies. If one of us is in need, the rest help out. And we don't steal from one another, ever" (Butler 2007: 301).[14] While Lauren occupies the moral center in Butler's novel, in *The Road* it is above all the boy who represents human compassion and the moral law. In one of the last conversations between father and son, the father prepares his son for his journey alone after his own death. The boy says he wants to come with him, but his father insists: "You cant. You have to carry the fire." When the boy expresses doubt and asks "Is it real? The fire?" he explains: "It's inside you. It was always there. I can see it" (299). The motif of "carrying the fire" is thus linked to how father and son represent the survival of good people.

However, considering the way in which both the plot structure and themes to a large extent repeat the patterns of light versus darkness, and

14. Also Butler's novel could be considered a road novel of sorts, but the band of migrants, formed en-route, end up in a place where they intend to establish their utopian community, based on tilling of the earth in an enclave away from a world that has gone wrong.

civilization versus savagery, the trope of "carrying the fire" takes on added meaning. The myth of Prometheus who brings fire and civilization to humans comes to mind. Likewise, the catastrophe of the novel can be likened to the evils that flew out of Pandora's box to plague humanity.[15] Set up against the dark forces of evil and chaos, the trope signifies a glint of hope for a better future. In the final pages, after the father's death, it is linked to the words spoken by the woman of the family that invite the boy to join them: "She said that the breath of God was his breath yet though it pass from man to man through all of time" (306). Thus, while the ending of the novel is open, allowing for speculations as to the fate both of the boy and his world, a spark of hope and redemption is contained in the almost messianic motif of "carrying the fire," and the theme of epic quest is retained.[16] Thereby, *The Road* is a critical dystopia as defined by Raffaela Baccolini and Tom Moylan: "the new critical dystopias allow both readers and protagonists to hope by resisting closure: the ambiguous, open endings of these novels maintain the utopian impulse within the work" (2003: 7).

6. Conclusion

As if adopting Lawrence Buell's "rhetoric of pastoral betrayal" (1998: 649), and drawing on structures borrowed from a range of literary modes

---

15. Both myths are recorded in Hesiod's *Works and Days*. *Encyclopædia Britannica* explains: "The Greek poet Hesiod related two principal legends concerning Prometheus. The first is that Zeus, the chief god, who had been tricked by Prometheus into accepting the bones and fat of sacrifice instead of the meat, hid fire from mortals. Prometheus, however, stole it and returned it to Earth once again. As the price of fire, and as punishment for humankind in general, Zeus created the woman Pandora and sent her down to Epimetheus (Hindsight), who, though warned by Prometheus, married her. Pandora took the great lid off the jar she carried, and evils, hard work, and disease flew out to plague humanity. Hope alone remained within" ("Prometheus" 2009: n.p.).
16. Charles M. Stang delivers a compelling commentary on the biblical allusions in the novel, centering on *logos*, *christos*, and *hodos* (road). It may be rewarding to explore possible Gnostic allusions in reference to certain passages in the novel, such as the notion of the divine spark hidden in the material body in reference to the motif of "carrying the fire" (e.g. 298), or Sophia in reference to the words spoken by the woman in the closing scene (306).

– the road novel, the parable, the anti-pastoral, and the critical dystopia – McCarthy's novel paints a frightening picture of a worst-case scenario of what the future may have in store for humanity and the physical environment. It may even be taken as a cautionary tale with regard to the direction in which our civilization appears to be headed, and to the extent that the text reads like a portentous warning, it verges on the jeremiad. In this respect, Lawrence Buell's view that toxic discourse "insists on the interdependence of ecocentric and anthropocentric values" (1998: 639) is worth emphasizing. "However one might wish otherwise," he argues, "the modern nature that toxic discourse recognizes as the physical environment humans actually inhabit is *not* a holistic spiritual or biotic economy but a network or networks within which, on the one hand, humans are biotically imbricated (like it or not) and, on the other hand, nature figures as modified (like it or not) by techne" (Lawrence Buell 1998: 657). Assuming that the catastrophe that devastated the world of McCarthy's novel is caused by human intervention, the environment is "modified ... by techne" to the extreme and transformed into an anti-landscape. Even when taking into account the allegorical form of the novel, one may still argue that it conveys an ecopoetic or ecodystopic vision or sensibility, which does not mourn the loss of pristine nature, but which serves to sharpen our awareness of dependence on the world we inhabit and are part of. In other words, although it is highly unlikely that the promotion of environmental concerns in any narrow sense motivated McCarthy to write *The Road*, the anti-landscape or "landscape of fear" of the novel – described according to the parameters of "toxic discourse" – may certainly move readers to such awareness.

At one level, McCarthy's novel can be read as an ecodystopian parable, which provides an *exemplum* regarding *oikos* in the inclusive sense of the word, which is to say, not how the individual should manage his or her personal household, but how humanity should manage their common household, the world that they are all part of. At another level, the text reads as an allegory that goes beyond a concern for the tangible word, or *ta physika*, and into the realm of *ta meta ta physika,* presenting a meditation on metaphysical concerns, in which it is possible to discern an inquiry into the limits of evil and the grotesque, a theodicy, and even a messianic quest.

Bibliography

Arnold, Edwin T. 1992. "Naming, Knowing and Nothingness: McCarthy's Moral Parables" in Arnold, Edwin T. and Dianne C. Luce (eds) *Cormac McCarthy*. Special issue of *Southern Quarterly* 30(4): 31-50.

Atwood, Margaret. 2010 (2009). *The Year of the Flood*. London: Virago.

Baccolini, Raffaella and Tom Moylan. 2003. "Introduction: Dystopia and Histories" in Baccolini and Moylan (eds) *Dark Horizons: Science Fiction and the Dystopian Imagination*. London: Routledge: 1-12.

Bakhtin, Mikhail M. 1990 (1937-38; 1973). "Forms of Time and of the Chronotope in the Novel: Notes toward a Historical Poetics" in Bakhtin, *The Dialogic Imagination: Four Essays* (ed. Michael Holquist, tr. Caryl Emerson and Michael Holquist). Austin: University of Texas Press: 84-258.

Balée, Susan. 2007. "Jim Crace's Violent Verities" in *Hudson Review* 60(3): 517-27.

Blackmore, Tim. 2009. "Life of War, Death of the Rest: The Shining Path of Cormac McCarthy's Thermonuclear America" in *Bulletin of Science, Technology and Society* 29(1): 18-36.

Buell, Frederick. 2003. *From Apocalypse to Way of Life: Environmental Crisis in the American Century*. New York: Routledge, 2004.

Buell, Lawrence. 1998. "Toxic Discourse" in *Critical Inquiry* 24(3): 639-65.

Buell, Lawrence. 2001. *Writing for an Endangered World: Literature, Culture, and the Environment in the U.S. and Beyond*. Cambridge, MA: Harvard University Press.

Butler, Octavia. 2007 (1993). *Parable of the Sower*. New York: Grand Central.

Carson, Rachel. 1963 (1962). "A Fable for Tomorrow" in Carson, *Silent Spring*. London: Hamish Hamilton: 3-4.

Frye, Steven. 2009. *Understanding Cormac McCarthy*. Columbia: University of South Carolina Press.

Gifford, Terry. 1999. *Pastoral* (The New Critical Idiom). London: Routledge.

Grammer, John M. 1992. "A Thing Against Which Time Will Not Prevail: Pastoral and History in Cormac McCarthy's South" in Arnold, Edwin T. and Dianne C. Luce (eds) *Cormac McCarthy*. Special issue of *Southern Quarterly* 30(4): 19-30.

Graulund, Rune. 2010. "Fulcrums and Borderlands: A Desert Reading of Cormac McCarthy's *The Road*" in *Orbis Litterarum* 65(1): 57-78.

Guillemin, Georg. 2004. *The Pastoral Vision of Cormac McCarthy*. College Station: Texas A and M University Press.

The Holy Bible, Containing the Old and New Testaments. 1989. New Revised Standard Version. Nashville: Thomas Nelson Publishers.

Jackson, John Brinckerhoff. 1984. *Discovering the Vernacular Landscape*. New Haven: Yale University Press.

Kerouac, Jack. 1976 (1957). *On the Road*. London: Penguin.

Kollin, Susan. 2011. "'Barren, silent, godless': Ecodisaster and the Post-Abundant Landscape in *The Road*" in Spurgeon, Sara L. (ed.) *Cormac McCarthy*: All the Pretty Horses, No Country for Old Men, The Road. London: Continuum: 157-71.

McCarthy, Cormac. 2006. *The Road*. London: Picador.

McCarthy, Cormac. 2010 (2006). *Sunset Limited: A Novel in a Dramatic Form*. London: Picador.

Monbiot, George. 2007. "Civilisation Ends with a Shutdown of Human Concern: Are We There Already?" in *Guardian* 30 Oct. On line at: http://www.guardian.co.uk/commentisfree/2007/oct/30/comment.books (consulted 13.01.2010).

Morgan, Wesley G. 2008. "The Route and Roots of *The Road*" in *Cormac McCarthy Journal* 6 (Autumn): 39-47.

Moylan, Tom. 2000. *Scraps of the Untainted Sky: Science Fiction, Utopia, Dystopia*. Boulder: Westview.

Phillips, Dana. 2011. "'He ought not have done it': McCarthy and Apocalypse" in Spurgeon, Sara L. (ed.) *Cormac McCarthy*: All the Pretty Horses, No Country for Old Men, The Road. London: Continuum: 172-88.

"Prometheus." 2009. *Encyclopædia Britannica Online*. On line at: http://www.britannica.com/EBchecked/topic/478684/Prometheus (consulted 24.05.2009).

Spurgeon, Sara L. 2011. Introduction in Spurgeon (ed.) *Cormac McCarthy*: All the Pretty Horses, No Country for Old Men, The Road. London: Continuum: 1-22.

Stang, Charles M. 2011. "Ash and Breath: Christ on 'The Road.'" Review of McCarthy (2006), and *The Road* (dir. John Hillcoat) in *Harvard Divinity Bulletin* 39(1-2). On line at: http://www.hds.harvard.edu/news-events/harvard-divinity-bulletin/articles/ash-and-breath-christ-on-the-road (consulted 20.08.2011).

Tuan, Yi-fu. 1980 (1979). *Landscapes of Fear*. Oxford: Blackwell.

Wood, James. 2007. "Getting to the End." Review of McCarthy (2006) in *The New Republic* 21 May. On line at: http://www.tnr.com/print/article/getting-the-end (consulted 10.10. 2010).

7

Anesthetic Landscapes: Reflections on the Photography of John Ganis

Hannes Bergthaller

This paper employs Wolfgang Welsch's concept of a "dialectic of the aesthetic" to discuss the landscape photography collected in John Ganis' *Consuming the American Landscape* (2003). According to Welsch, the modern drive towards aestheticization of the life world, in a dialectical reversal, produces anaesthetization. Ganis' photographs draw attention to the sharp disjunction between the aesthetic effect of a landscape and its ecological condition, and deliver a subtle critique of traditional landscape aesthetics, which both facilitated the commodification of landscape and anaesthetized people to the latter's consequences. In doing so, they present a plea for ecological education and provide a visual record of what collective human life on the land actually looks like today.

Keywords: environmental education; landscape aesthetics; photography; Welsch, Wolfgang

1. After the Substantive Landscape

"Landscape" is notoriously ambiguous word. It can refer to *territory* – a bounded geographical area – as well as to *scenery*, i.e. to the visual aspect that a territory presents to an observer, or to the *representation* of scenery in the medium of painting or photography. The potential confusion between these meanings has lead some geographers to abandon the concept altogether (Olwig 630). As Kenneth Olwig has argued, however, it is precisely this fuzziness that makes it so useful: it forces us to attend to the *interplay* between different semantic registers which, if grasped in its historical specificity, allows for an adequate understanding of the ways in which human communities are invested in their physical environments. In their original sense, the terms "landschap" (Dutch), "Landschaft" (German), or “landskaber” (Danish) denoted not only a geographical area, but also the body politic incorporating the territory in question; a “land” was constituted not merely by its physical features, but principally by the set of customary laws that bound together its inhabitants. When the genre of landscape painting gained popularity in Northern Europe in the course of the sixteenth century, this development reflected the interest of the rising mercantile class in recovering and protecting the traditions of local autonomy and self-government that had come under threat from the emerging absolutist state (Olwig 631-35). In depicting scenery, early landscape painters such as Pieter Breughel the Elder simultaneously dramatized the customary use rights and obligations that tied communities to their land (Rosenthal 12). Implicit in the famous *Hunters in the Snow* (1565; illustration 1), for example, is the right of free peasants to hunt and gather firewood in the forest, as well as the function of the frozen pond as a communal gathering place. The territorial and the aesthetic meanings of landscape are here inextricably entwined: the legal, political, and economic order of human communities is etched into a physical landscape; visual representations of that landscape at once reflect these social inscriptions and enhance their legibility.

Illustration 1: Hunters in the Snow

However, it is far from clear whether such a "substantive" concept of landscape, as neither simply territory or scenery, but rather "a nexus of community, justice, nature, and environmental equity" (Olwig 630-31), is still viable and relevant in the contemporary world, where the institutional and symbolic networks that constitute the social are much more spread out and difficult to project onto a single geographical plane. This challenge is particularly acute with regard to "anti-landscape": while the destruction of local environments immediately raises ethical, political, and legal questions, it often seems to occur precisely because those most responsible for an environment's destruction are not imaginatively invested in it – that is to say, because they have not conceptualized it as a "substantive" landscape. Against this background, it must be asked what role traditional forms of landscape representation can still play in addressing the set of problems denoted by the term "anti-landscape." In this essay, I pursue this question by engaging with John Ganis' photography book *Consuming the American Landscape* – a work particularly well-suited for this purpose because it not only seeks to register the consequences of society's refusal to recognize itself in the landscapes it engenders, but may also be seen as investigating the possibility of resuscitating the kind of communal bond to the land suggested by earlier forms of landscape representation. In doing so, I argue, Ganis' photographic art probes the limitations that visual representations of landscape invariably run up against when they seek to illuminate contemporary problems of ecological degradation. It delivers a subtle critique of traditional American landscape iconography, which aestheticized the land in a manner that facilitated its commodification even as it desensitized people to the consequences of the latter. *Consuming the American Landscape* thus draws attention to and seeks to come to terms with what the German philosopher Wolfgang Welsch has described as a "Dialektik des Ästhetischen" (15), whereby the drive towards *aesthetization* of the lifeworld produces social *anesthetization*.

2. Consuming the American Landscape

Published in 2003, *Consuming the American Landscape* comprises 80 photographs by John Ganis, together with six untitled poems by the anthropologist Stanley Diamond and two brief essays by the curator Robert Sobieszek and the landscape architect George F. Thompson. The

Illustration 2: Earthmover, Texarkana, Texas

photographs are grouped into loose thematic clusters, all except the first prefaced by a poem: a section with pictures of land being cleared for development (12 images; illustration 2) blends into one about oil drilling (4 images; illustration 3), followed by sections on landfills and waste disposal (9 images; illustration 4), mining (22 images; illustration 5), housing developments (15 images), and logging (15 images). What is perhaps most striking about these images, at first sight, is the reticence of their visual language. The steady, impassive gaze that they direct at the land equally enfolds fir trees and electricity towers, rocks and barrels, hubcaps and bushes, the curvature of hills and roads, as elements of a unified, carefully balanced composition. These are landscapes that cannot readily be resolved into the dichotomies of natural *versus* artificial, or wild *versus* cultivated. Although they are impure, even defiled, they have an undeniable aesthetic appeal which, the unobtrusive formalism of the photographs seems to suggest, is not artfully imposed but springs from the landscapes themselves.

What the images lack in pathos, however, is more than compensated for by the accompanying poems and Thompson's concluding essay. Diamond inscribes the photographs into a starkly declensionist narrative. His second

Illustration 3: All Terrain Vehicle Damage, Peters Creek, Alaska

Illustration 4: Landfill Interior, Michigan

poem contrasts the cultures of Native Americans, materially and spiritually in tune with the land, as their way of life does not leave any lasting traces ("Beads of leather / And shells / Fibers of Baskets / And bows / [...] Real enough to disappear / Into origins / And Overall / The imprint / Of aspiring wonder / [...] Earth returning to Earth / Say the Elders / Now and forever", 43), to the suicidal wastefulness of industrial modernity ("What has been absolute / for forty-thousand years / The ground of being itself / The pact we make with infants / No longer holds / And over all / The imprint / Of killing machines / At which / Only the roach will one day wonder", ibid.). The subsequent poems lament the corrosion of rural communities by the mining industry ("In the beginning the mine / was on the edge of town / In the end, the town / was on the edge of the mine", 55), the disappearance of traditional farming and its ethos of sustainable cultivation ("The small farms are gone / And with them / The last tie to the land", 87), and the destructive logic of capitalism ("The dream of profit / is a bad dream / Of flat horizons / and cracked cement", 109). Diamond's openly didactic poems are concerned precisely with the loss of a substantive relationship to the American landscape – with the failure of modern Americans to see the devastation which their way of life has wrought on the land, and to understand its ecological consequences. The images in *Consuming*

Illustration 5: Remediated Mine Tailing, Sunny Side, Utah

the American Landscape, Diamond implies, force us to face up to these consequences and to mend our ways. This is the argument spelled out in Thompson's essay "Connecting the Dots." The landscapes depicted in Ganis' work, he tells the reader, "lay before us as evidence of our deeds, our aspirations, our shortcomings. [...] They direct us to overcome the guilt of seeing land mistreated in the interests of self-indulgence; they implore us to create better places in which to live, work, and explore." (137-38)

With reference to the terms introduced at the outset of this essay, one may summarize Diamond and Thompson by saying that Americans have allowed their land to be destroyed because they have failed to properly correlate the aesthetic and the territorial dimensions of the American landscape, to read the visible surface of the land as an inscription of their social order. Ganis' photographs could thus be understood as representing anti-landscapes in a double sense: they depict places (1) that are ecologically damaged so that their ability to support life is diminished, and (2) whose capacity to stand as metonyms of the social order is disavowed. They bespeak a society in which the relationship to the land is depersonalized, its use entrusted not to local communities but to corporate owners; a society that does not look *after* its landscapes because it does not really look *at* them – a circumstance underscored by the fact that (with

Illustration 6: Land Development, Florida

the exception of a single photograph to which I will return below) they are entirely devoid of people. The promise of Ganis' art, then, would be the recovery of a substantive understanding of the American landscape – a restoration of the broken link between the American polity and the physical landscapes sustaining it. It should teach Americans to read their landscapes, once more, as signs of national character.

3. Representing the Anti-Landscape

Appealing though such an approach to Ganis' photographs may be, I wish to argue that it misses much of their critical thrust. Thompson's confidence in their moral and political efficacy rests on their supposed ability to capture the reality of what they represent: "Landscapes and places are tangible realities, they don't lie." (137) In this view, the rhetorical power of *Consuming the American Landscape* is ascribed to the landscapes themselves, more than to Ganis' photographic art: "John Ganis did not seek out these landscapes […], but *they found him* through his lens. *They showed him* that there is a direct relationship between one's lifestyle and one's national landscape […]." (138; my italics) While one may wish to dismiss these statements as instances of purely rhetorical hyperbole, they lead straight to the heart of the aesthetic problems posed by the notion of anti-landscape.

Underlying Thompson's account is the assumption that environmental damage is a fact immediately available to our senses, and therefore amenable to representation in the traditional mode of landscape photography that attempts to capture (and idealize) the visual impression which a particular piece of scenery leaves on the viewer. In a very basic sense, Ganis' photographs can be said to bear this out: they show land that has been leveled and cleared by earth movers (illustration 2); torn up by excavators (illustration 6), dirt bikes (illustration 7) and ATVs (illustration 3); disfigured by landfills (illustration 4) or mining operations (illustration 5, 8). Clearly, these are all visual images of "tangible" places that have been damaged, more or less severely, by human activity. Yet even if one grants that these places "don't lie," as Thompson says, one would hardly wish to claim that they speak in a particularly forthright manner. In each of these photographs, the damage is of a qualitatively different kind, and what is striking about them is not so much the visible impact of human activities on the land as the disconnect between the latter and the severity of environmental damage. The terraced mine tailings in the picture "Remediated Mine Tailings, Sunny Side, Utah" (illustration 5), with their vague resemblance to a Maya pyramid, constitute a visually dramatic intrusion on the landscape; however, without further information, we really have no warrant for assuming that the place will not return to an ecologically healthy state, in the sense of being able to support a biologically diverse community, within a reasonably short time. The photograph shows a landscape that has been altered by human activity, but not necessarily one that has been ecologically diminished. The body of water depicted in figure 8, on the other hand, is explicitly identified by the subscript as a "Toxic Coal Mining Pond." The image shows a place that has in all likelihood become uninhabitable for humans – yet apart from the bleached appearance of the trees and rocks at the center of the photograph, which might be attributable to ecologically harmless chalk dust, there is nothing to disrupt the idyllic appearance of the scene. This may very well be a photograph of an anti-landscape in the strictest sense, but the "anti" of this landscape is not part of the picture, as it were.

Illustration 7: Dirt Bike Trails, California

Illustration 8: Toxic Coal Mining Pond, Pennsylvania

Rather than simply showing us environmental destruction, Ganis' photographs thus draw our attention to the sharp disjunction between the aesthetic effect of a landscape and its ecological condition. What we see is clearly not what we get. One can understand this as a version of the problem that Aldo Leopold already described in *A Sand County Almanac* when he wrote: "[O]ne of the penalties of an ecological education is that one lives alone in a world of wounds. Much of the damage inflicted on the land is quite invisible to the layman. An ecologist [...] must be the doctor who sees the marks of death in a community that believes itself well and does not want to be told otherwise" (197). If anti-landscapes are defined by their diminished capacity to support life, then they defy the traditional forms of landscape representation, because ecological health is a feature of the land that is accessible neither to the naked eye nor to the camera lens. If, for example, we want to get a grasp of the environmental realities at the Hanford Reach (illustration 9) – a stretch of the Columbia River that was for several decades affected by periodical releases of highly radioactive cooling water from the nearby nuclear reactor – a photographic image will not take us very far. Of course, even a landscape like the one depicted in Breughel's "Hunters in the Snow" is not legible without knowledge about

Illustration 9: The Hanford Reach

the social order to which it was coupled; yet the clues required in order to establish this linkage are all readily available to the unaided eye. The Hanford Reach, on the other hand, presents a landscape that cannot be properly understood at all without reference to parameters that can only be grasped with the help of advanced measuring equipment. Any serious attempt at a visual rendering of an anti-landscape of this kind would therefore have to resort to entirely different modes of representation, as illustrated by, for example, *Toxmap*, a hyperlinked map of superfund sites in the US provided by the National Library of Medicine (illustration 10). It would have to call on what Ursula Heise, following media theorist Lev Manovich, refers to as a "database aesthetic" (67). (See page 120.)

4. Anti-Landscape and the Dialectic of the Aesthetic

Anti-landscapes are thus an exemplary instance of what the sociologist Ulrich Beck has termed the "the expropriation of the senses" (55): the

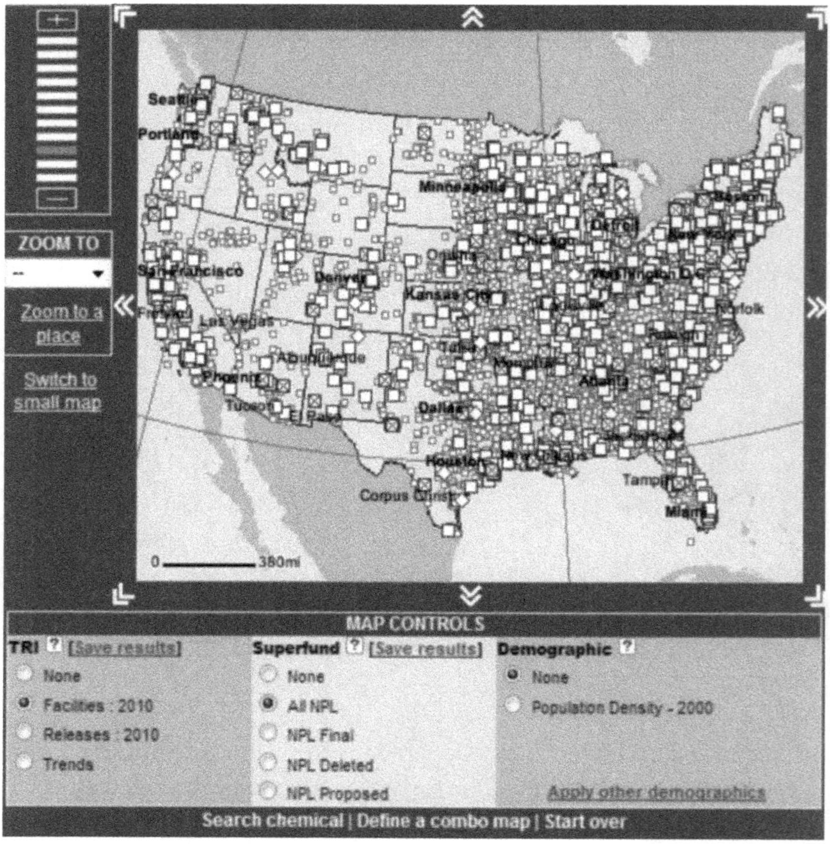

Illustration 10: Toxmap

advanced technologies of modern society have created new risks that lie beyond the ken of humans' natural sensorium, and they have made calculable already existing risks formerly unknown to us. In either case, advanced technology has drastically undercut the authority of natural sensory perception. That there are things we cannot see, hear, smell, or feel, but which are nonetheless real is, of course, something that people have always known, as the history of the world's religions and the long predominance of metaphysics as a branch of philosophy amply show. What is new about the modern situation is the degree to which our senses can no longer be trusted "even in their proper domain" (Welsch 18; all translations are mine). This is a situation perhaps most familiar with regard to food, where we have become quite used to being suspicious of the evidence from our own senses, and instead rely on nutrition charts, official certifications and statistical evidence. As Wolfgang Welsch puts it: "trust in our bodily senses [...] is not just antiquated, but has become a trap. Things that please also destroy us. Technology has transformed reality [...] to such an extent that our inert, conservative (in the sense that they are calibrated to a nature that is increasingly disappearing) senses have become not just unreliable, but counter-productive – they are now agents of the false" (19).

According to Welsch, this development should be seen in the larger context of a process he describes as the "dialectic of the aesthetic," in analogy to Adorno and Horkheimer's well-known "dialectic of enlightenment." Ever since the formation of aesthetics as a philosophical discipline in the mid-eighteenth century, the idea of sensitizing humanity to aesthetic values and transforming reality with reference to the latter was central to the project of modernity (28f). However, just as the progress of enlightenment engendered new and improved forms of barbarism, so increasing *aesthetization* led, in a dialectical reversal, to increasing *anesthetization*. Every attempt to ameliorate society in terms of an aesthetic ideal led to a situation in which that ideal itself became oppressive, blunting sensibilities instead of enhancing them. Hence the history of the arts in the modern era can be written as a history of rebellions against aesthetic schematizations that are rejected because they have, through habituation, become *anesthetic*, in the sense of obstructing authentic perception. Originating in a critical reaction to the horrors of early industrialization, the Romantic idealization of nature gradually turned into an anodyne that only made it easier to accommodate oneself to what seemed to be the inevitable. It was overturned by modernists of various stripes, all of whom tried, in one way or another, to inoculate their audience against the aesthetic appeal of Romantic platitudes by adopting strategies of *purposeful* anesthetization – an artistic program easily recognizable in works as different as the poetry of T. S. Eliot ("Let us go then, you and I / When the evening is spread out against the sky / Like a patient etherized upon a table"; 2289), Marcel Duchamp's comments on his "readymades," or in the famous sliced eyeball-scene from Luis Buñuel's *Un Chien Andalu* (Welsch 37). But even the surrealists' paradoxical attempts to *re*-sensitize viewers by *de*-sensitizing them, i.e. by undercutting the appeal of established aesthetic schemata and exposing their anesthetic effects, did not escape the "dialectic of the aesthetic" – they were absorbed into the visual language of advertising, becoming themselves habituated and in effect anesthetic.

The "dialectic of the aesthetic" thus denotes a process in which every movement towards aestheticization entails a moment of the anesthetic, triggering a countervailing movement for which the same holds true. For Welsch, this dialectic has its roots in the fundamental fact that every act of sensory perception is possible only on the basis of exclusions that are constitutive to it. This is illustrated by the blind spot in our retina, a simple example that can be generalized to those higher-order schematizations of perception which concerned the gestalt psychologists and that, on yet another level, are the stuff of art history. Welsch's summary must be understood as applying to all levels of sensory perception: "We don't see because we are not blind, but we see *because* we are blind to most things; accordingly, any act of making something visible simultaneously makes something else invisible. No *aisthesis* without *anaisthesis* [...]" (31-32).

What this suggests with regard to the photography of John Ganis is that if we wish to understand the specific convergence of the aesthetic and the political by which they acquire their rhetorical force, we need to examine it not only in relation to the forms of environmental destruction which the concept of anti-landscape most immediately calls up, but also within the iconographic tradition that it both continues and disrupts. From this perspective, the disjunction between ecological reality and aesthetic perception which anti-landscapes can be said to exacerbate is only the latest version of a much more general problematic, namely the complex interplay of revelation and occlusion that constitutes the "dialectic of the aesthetic." Ganis' photographs pursue a strategy of anaesthetization in Welsch's sense: they expose the anesthetic effect of the aesthetic schemata that have traditionally governed the representation of landscape and thus seek to inoculate us against their seductive appeal. If this were not somewhat at odds with their determined refusal of histrionics, one might be tempted to say that they repeat for the genre of landscape photography what Buñuel's sliced eyeball did for the cinema.

5. Representing the Actual Landscape

We have thus arrived at an interpretation of Ganis' photography that is almost diametrically opposed to the one suggested by the accompanying texts in *Consuming the American Landscape*. Recall that Thompson views them as a photographic jeremiad against American consumerism and attributes their critical force to the fact that they are transparent representations of ecological destruction. This assumption is problematic insofar as ecological health is a feature of the land that largely eludes natural human vision. What the photographs dramatize is therefore not ecological destruction itself, but rather the limitations which traditional modes for the representation of landscape encounter when they seek to depict that destruction. If the photographs are able to produce a sense of guilt in the viewer, as Thompson argues, it is not because they faithfully document the process of landscapes degenerating into anti-landscapes, but because they refer to a familiar tradition of landscape iconography whose generic conventions they purposefully violate.

If we limit our purview to the American scene, this iconographic tradition can be said to stretch back to nineteenth century landscape photographers such as Timothy O'Sullivan or William Henry Jackson and, before that, to the paintings of the Hudson River school. If any particular photographer can be singled out whose work has ensured the continuing vitality of this tradition, it must be Ansel Adams, "the Great Oedipal Father of American Landscape Photography," in Rebecca Solnit's words (91), whose black-and-white photographs of the American West have lost little of their popularity since the 1940s when they first began to circulate through the mass media. It was during this time that Adams turned from the modernism of his earlier work as a member of the f/64 group, which had favored chillingly abstract compositions and was aimed at a small audience of connoisseurs, towards a more dramatic and sentimental style that presented the American landscape in grand panoramic vistas, as exemplified by his famous picture "The Tetons and the Snake River" (illustration 11). Through coffee table books and calendars, these images were made available to a wide popular audience, and, particularly after Adams began to work closely with the Sierra Club, achieved iconic status for the burgeoning environmental movement in the US.

It is this tradition of explicitly environmentalist photography that *Consuming the American Landscape* latches on to; its very formal organization recalls books such as *This is the American Earth* (published in 1960 and generally regarded as the first coffee table book), which paired the work of Adams and several other well-known photographers with short prose pieces by art critic Nancy Newhall. As politically effective as these

Illustration 11: The Tetons and the Snake River

images may have been in their time, they also inadvertently contributed to the very process that they were meant to oppose, thus falling prey to the "dialectic of the aesthetic": By stoking the desire for a life closer to the kind of natural beauty enshrined in *This is the American Earth*, they did their share to fuel the population boom experienced by the Western states of the US following WWII. Not coincidentally, car companies soon became (and remain until today) fluent in the visual language of Adams' landscape photography. The very prevalence of such images, which (following Adams' example) carefully erased the signs of human presence, taught people to overlook the actual impact of development on the land.

It may be more than just a pun, then, to say that *This is the American Earth* helped to establish the iconographic parameters for the consumption of the American landscape. The images in *Consuming the American Landscape* thus not only document the process denoted by its title, but also reflect on this complicity between the aesthetization of landscape in the iconographic tradition represented by Adams and the dissolution of the same landscape into an ever-expanding sea of landfills and suburban sprawl. The photograph which makes this point with the greatest clarity is the title image of the book (illustration 12). It shows a panoramic view of South Dakota's Badlands National Park which could just as easily figure in a coffee table book published by the Sierra Club, were it not for the conspicuously inconspicuous piece of piping that juts out of the cliff-side in the lower half of the picture. At the center of the photograph is a small family, dwarfed by the surrounding magnificence. They have come, we may safely assume, to look at an Adams landscape. Facing away from the camera, they are also facing away from that piece of piping that stands metonymically for the infrastructure that has made their sojourn in this park possible, in the first place.

The proper response to Ganis' photographs, then, is neither simply a sense of guilt, nor of outrage, nor an emphatic call to renew the substantial bond between the people and the land they inhabit (desirable as that may be), but a much more complicated mix of thoughts and emotions. These images seek to elicit in the viewer partly an acceptance of the inevitable impact that human inhabitation has on the land, and partly an acknowledgement of the need for an "ecological education" (in Aldo Leopold's sense) that

Illustration 12: Family at Badlands National Park, South Dakota

goes beyond the cult of wilderness and its logical extension, the lament for the raped virgin land (as it was characteristic for the photographers of the momentous *New Topographics* show of 1975, whose work can be seen as an immediate forerunner to that of Ganis; cf. Solnit 92-94). If their mood is elegiac, then what they mourn is not only the demise of the American wilderness, but also the obsolescence of traditional landscape photography itself, which has finally been stripped of its pretensions to represent an environment that must be understood as encompassing not only mountains and rivers, but also international stock markets, invisible toxins, and illegal immigrants, and whose phenomenology therefore explodes the visual bounds of the scenic. *Consuming the American Landscape* does not so much depict anti-landscapes that have lost their capacity to support collective human life as it deflates an iconography that has become anesthetic. In doing so, it shows us what collective human life on the land actually *looks* like today – no more, and no less.

Bibliography

Adams, Ansel and Nancy Newhall. 1960. *This Is the American Earth*. Sierra Club.

Beck, Ulrich. 1999. *World Risk Society*. Polity Press.

Eliot, T. S. 1915. "The Love Song of J. Alfred Prufrock." *The Norton Anthology of English Literature*, vol.2, 8th ed. 2289-93.

Ganis, John. 2003. *Consuming the American Landscape*. Stockport, UK: Dewi Lewis.

Heise, Ursula. 2008. *Sense of Place and Sense of Planet. The Environmental Imagination of the Global*. Oxford: Oxford University Press.

Leopold, Aldo. 1970. *A Sand County Almanac. With Essays on Conservation from Round River.* 1949. New York: Ballantine.

Olwig, Kenneth R. 1996. "Recovering the Substantive Nature of Landscape." *Annals of the Association of American Geographers* 86.4, 630-653.

Rosenthal, Michael. 1982. *British Landscape Painting*. London: Phaidon.

Solnit, Rebecca. 2001. *As Eve Said to the Serpent. On Landscape, Gender, and Art.* Athens, GA: University of Georgia Press.

Thompson, George F. 2003. "Connecting the Dots." John Ganis, *Consuming the American Landscape.* Stockport, UK: Dewi Lewis, 137-38.

Toxmap. 2012. The United States National Library of Medicine. <http://toxmap.nlm.nih.gov/toxmap/main/index.jsp>. April 15.

Welsch, Wolfgang. 1990. *Ästhetisches Denken*. Stuttgart: Reclam, 1990.

8

Fear and Fascination: Anti-Landscapes between Material Resistance and Material Transcendence

Werner Bigell

This article examines the mechanisms that turn landscapes into anti-landscapes: projection, material interference, and ideological contradiction. Landscape and anti-landscape are dialectical twins; whereas landscape affirms cultural and aesthetic values, anti-landscape negates them through material resistance. This negation creates a sense of material transcendence, the aesthetic appeal through understanding that the world is larger than the projections onto it. Negation and material transcendence is the common denominator for anti-landscapes such as the exclusion zone around Chernobyl and the "feral houses" of the American suburb.

Keywords: "American Beauty"; Chernobyl; Ecotopia; Filatova, Elena; Lem, Stanislav; suburbia; Tarkowsky, Andrei; Zizek, Slavoj

1. Introduction

J. B. Jackson's statement that landscape is defined by the human presence in it appears to be anachronistic in an age of environmental awareness, with its fascination for the wild, yet it highlights the fact that meaning is not inherent in space but is projected onto it. Projection is a two-step process: when the Puritans came to America, they first projected the image of a New Israel onto space and then later sought to remake that space into the City upon the Hill. Landscape is the result of a cultural projection transforming physical space. Projection is not limited to humans; in his story "The Encantadas," Herman Melville describes how tortoises struggle incessantly against rocks in their path rather than going around them because of their "impulse to straightforwardness in a belittered world" (1856). The tortoises project straight lines onto their environment, and their projections transform the land, creating "deep ruts by ages and ages of the slow draggings." Projection is a biological mechanism transforming the environment, but only human beings transform space into landscapes, which are invested with cultural meaning through projections of a scientific, economic, aesthetic, political, or spiritual character.

Projection is preceded by representation. Representation can refer to the transfer of material and spatial patterns onto media such as maps, and it also refers to the formation of images of the external world in the mind. The aim of image formation is survival, not creating a mirror of the external world. Richard Dawkins describes the relation of image and reality from a biological viewpoint: "we animals inhabit a virtual world, constructed from elements that are, at successively higher levels, useful for representing the real world – which is exactly as it should be if our constrained virtual reality software is any good" (1998, 275). The linguist Stephen Pinker argues in a similar vein: "We are organisms, not angels, and our minds are organs, not pipelines to the truth. Our minds evolved by natural selection to solve problems that were life-and-death matters to our ancestors, not to commune with correctness" (1997, 561). Dawkins and Pinker argue against direct mental representation of the material world, but their stripe of constructivism allows a close functional relation of mind and matter because images are made for engagement in the world. Also the sociologist Norbert Elias (1986) describes a functional relation of mind and material world and distinguishes between external reality, knowledge, and the subject. Since knowledge is part of a survival strategy, it has to be "reality congruent." Humans are engaged in the material world and project mental images onto it, transferring cultural meaning onto space. Projection may alter a space materially or not, but it always alters the conception of it. A functional relation between representation, projection, and engagement also implies the possibility of incongruence and failure.

In society, mental images are projected onto the world not primarily for survival but as an expression of individual and collective engagement in the world. When this process fails, and the failure manifests itself in space, the result is dysfunctional, and quite likely an anti-landscape. One can see projection failures as part of biological and cultural adaptation to the world, a call to not confuse image and material reality and to learn from failure to create a better image; negation here is a corrective, not a predicament. Incongruence and failure have an aesthetic component, as negation is a confirmation of materiality. In the act of touching there is limitation of movement. Yi-Fu Tuan states, "[t]ouch is the direct experience of resistance, the direct experience of the world as a system of resistance and pressures that persuade us of the existence of a reality independent of our imaginings" (1990, 8). Metaphorically speaking, the highlighted materiality of the anti-landscape is "touching," which is attractive for a culture that values authenticity. It is important to note the difference between this experienced-based and biological definition of the relationship of people to their surroundings with the classical definitions of landscape. Tim Cresswell argues that traditionally "[l]andscape is an intensely visual idea" and that "in most definitions of landscape the viewer is outside of it" (2004, 10). The dominance of the visual in landscape is caused by a class-bias (the material engagement of the farmer plays a lesser role than the visual consumption of the privileged visitor) as well as an idealist bias, valuing ideas more highly than direct contact with the material world. The negation in the encounter with the anti-landscape could be a vehicle to re-infuse a sense of materiality and limitation into our perception of landscape.

Negation presents itself in four different forms. First, there is the case where negative cultural meaning is projected onto landscape, such as the depiction of the desert in the Bible. Second, a cultural projection fails when it clashes with material reality. An example here is projection of the controllability of atomic power onto Chernobyl. Third, projections fail because the spaces they are projected onto are non-responsive. The Polish science-fiction author Stanislaw Lem creates worlds that remain radically other, showing that what seem to be "universal" human ideas are merely terrestrial in the universe. Fourth, projections can fail because of inherent ideological contradictions. An example is the American-style suburb, which negates both the city and countryside that it aims to reconcile. Yet paradoxically, all those failures can be appealing. Lem's non-responsive fictional worlds fascinate readers, the desert is a tourist destination, and "dark tourists" now explore both the exclusion zone of Chernobyl and the "feral houses" of abandoned suburbs.

2. The anti-landscape of negative projection

Landscapes do not have to be physically altered in order to be meaningful. Whereas the Alps had been a dangerous obstacle for most of European history, romanticism reevaluated mountain ranges as havens of purity over the valleys of industrial corruption. Also deserts and wilderness areas have had shifting meanings projected onto them: the desert of Sinai is the prototypical anti-landscape, a testing ground for the people of Israel and a space to dispose unwanted elements: "But with many of them God was not well pleased: for they were overthrown in the wilderness" (1 Corinthians 10:5). The desert for Jesus was both a place of temptation by the devil and for reaffirming his belief: "And he withdrew himself into the wilderness and prayed" (Luke 5:16). Anti-landscapes here are meaningful through negative projection but remain themselves unaltered.

Our age has turned many anti-landscapes into landscapes, but negativity lurks under the surface, as in the film "Deliverance" where a green landscape turns into an anti-landscape of horror. Often negativity is projected onto the city; the film "Taxi Driver" depicts New York as a place "where all the animals come out at night." Another trend in Western culture is to project ambivalent anti-landscapes onto the non-Western world. Anti-landscapes have migrated to the dangerous Orient (with its secret pleasures), claustrophobic and caste-ridden India (with its spirituality), the tropical heart of darkness (with its promise of promiscuity), or lawless Mexico (a place to party or to hide). Landscapes and anti-landscapes are dialectical twins, their opposition being fueled by the projection of contradictory desires.

3. Non-responsive anti-landscapes

The hidden worlds of the universe have been white screens for the projections of science fiction, but they can also be mirrors. In his novel *Solaris*, Stanislaw Lem illustrates the troubling possibility of an inhuman world. The planet Solaris is covered by an intelligent ocean as its sole inhabitant, but generations of researchers have only managed to describe phenomena on its surface in geocentric terms while failing to establish contact. Now interest in Solaris has subsided, and only one station is left in the planet's orbit. The three researchers there are challenged by a phenomenon that they are unable to report, for fear of being classified as mentally disturbed. The ocean creates apparitions of human beings, modeled on the memories of the researchers: it creates a mirror but remains itself incomprehensible. One of the researchers realizes that mirroring is at the center of their mission:

> We take off into the cosmos, ready for anything: for solitude, for hardship, for exhaustion, death. Modesty forbids us to say so, but there are times when we think pretty well of ourselves. And yet, if we examine it more closely, our enthusiasm turns out to be all sham. We don't want to conquer the cosmos, we simply want to extend the boundaries of Earth to the frontiers of cosmos. For us, such and such a planet is as arid as the Sahara, another as frozen as the North Pole, yet another as lush as the Amazon basin. We are humanitarian and chivalrous; we don't want to enslave other races, we simply want to bequeath them our values and take over their heritage in exchange. We think of ourselves as the Knights of the Holy Contact. This is another lie. We are only seeking Man. We have no need for other worlds. We need mirrors. We don't know what to do with other worlds. A single world, our own, suffices us; but we can't accept it for what it is. We are searching for an ideal image of our own world [...] (2003, 75-76)

The researchers remain caught in their human way of perception and try to rid themselves of the apparitions, unable to interpret them as contact to an intelligence that does not distinguish between self and other, past and present, idea and material reality. Many of Lem's fictional worlds portray anti-landscapes frustrating the human will to project a terrestrial "universal" onto the universe.

Anti-landscapes are spaces of material transcendence, where the limits of projection are manifest. They are suggestive of an underlying materiality. Edward Abbey saw the aesthetic potential in material transcendence when he described why he went into the desert:

> Near the summit I found an arrow sign, three feet long, formed of stones and pointing off into the north, toward the those same old purple vistas, so grand, immense, and mysterious, of more canyons, more mesas and plateaus, more mountains, more cloud-dappled sun-spangled leagues of desert sand and desert rock, under the same old wide and aching sky.
>
> The arrow pointed into the north. But what was it pointing at? I looked at the sign closely and saw that those dark, desert-varnished stones had been in place for a long, long time; [...] I studied the scene with care, looking for an ancient Indian ruin, a significant cairn, perhaps an abandoned mine, a hidden treasure of some inconceivable wealth, the mother of all mother lodes ... But there was nothing out there. Nothing at all. Nothing but the desert. Nothing but the silent world. That's why. (1991, 21-22, Abbey's emphasis)

Abbey does not invest the desert with negative or positive meaning but sees it as a non-responsive silent black screen pointing only at its materiality. The desert is appealing because it transcends projection and the aim of projection, usefulness: it is literally a wasteland. Slavoj Žižek discusses the aesthetic potential of waste: "[T]he properly aesthetic attitude of the radical ecologist is not that of admiring or longing for a pristine nature of virgin forests and clear sky, but rather that of accepting waste as such, of discovering the aesthetic potential of waste, of decay, of the inertia of rotten material which serves no purpose" (2011, 35). Anti-landscapes appeal because of their wastefulness, highlighting a materiality that serves no purpose.

Andrei Tarkovsky's film "Stalker" illustrates that the world appears to be non-responsive if one refuses to see it. The stalker guides two clients, a writer and a scientist, into the forbidden zone, to a place where wishes are fulfilled. The zone is an overgrown wasteland, full of deadly traps and industrial debris, and it is impossible to return the same way one came. For the Stalker the zone is home, whereas for the writer and the scientist it is only an obstacle for wish fulfillment. However, when they reach their destination, they are confused because they realize that their innermost wishes are anxiety-laden and elusive. They do not realize that the zone itself is a metaphor for life: it is dangerous, unidirectional, and wasteful. There is a spiritual tone in the film, but as Žižek points out, "for Tarkovsky, the very material process of corruption (decay, decomposition, rotting, inertia) is spiritual" and speaks of "Tarkovsky's spiritual materialism" (2011, 7). Whereas the visitors seek immaterial transcendence, the essence of the zone is waste. Waste is appealing because it escapes the utilitarian frames of modern civilization and of a romanticized nature where everything has a purpose in the web of life.

4. The physical resistance of the anti-landscape

Projection onto the material world involves abstraction. When constructing a house, predictions are made, based on abstractions about building materials, ground, climate, etc., and the projection may fail if there is an incongruence between abstraction and the material world. Scientific projection predicts the effects of the transformation of the material world but is limited by an increasing complexity, which reduces calculability. Whereas in an age of rapid industrial development and abundant resources the materiality of the world tended to be overlooked, the current environmental crisis is changing this, as Žižek argues: "[O]ver the last few decades, it has been fashionable to talk about the predominant role of 'intellectual labor' in our post-industrial societies – however, materiality is now reasserting itself with a vengeance in all its aspects, from the forthcoming struggles over scare resources (food, water, energy, minerals) to environmental pollution" (2011, 330).

The increasing control of the material world has lead to blindness towards material limits. Since industrialization has created large-scale projections, their failures have large-scale consequences. When cave dwellers dug a cave too wide, the collapsing roof may have killed a family, but when today earthquake and tsunami danger are miscalculated, the results can wipe out entire cities. Today the complexity of the material world is matched by a growing complexity of the social (including economic) world, also characterized by limitation of control. As Karl Marx observes: "A similar movement is going on before our own eyes. Modern bourgeois society, with its relations of production, of exchange and of property, a society that has conjured up such gigantic means of production and of exchange, is like the sorcerer who is no longer able to control the powers of the nether world whom he has called up by his spells" (1848). Today the powers of the nether world, social and productive forces, are accompanied by an increasing material incongruence, environmental and economic crises.

In his examination of the history of blackouts, David E. Nye shows that electrical light is a powerful sign of modern control, with the night being "treated as a canvas, which society may paint with lights" (2007, 77). Blackouts then can be seen as "a form of negation" revealing "an underlying disorder" (2007, 83). Hiding an underlying disorder can also be associated with natural light, as Paul Bowles shows in the novel *The Sheltering Sky* describing the blue desert sky: "the sky hides the night behind it, shelters the person beneath from the horror that lies above" (1990, 328). Material transcendence, evoking both fear and fascination, are associated with the absence of light. Electricity for light is produced in power plants, and when the projection of the behavior of the nuclear reaction in Chernobyl failed, it created a gigantic anti-landscape, an exclusion zone with uninhabited cities and villages. But if projection failed, the material world has not disappeared. Chernobyl today is a tourist destination, a prime example of "dark tourism," with several tour operators offering trips to the reactor and the abandoned city of Pripyat. In her online photo essays Elena Filatova explores the attraction of the exclusion zones in the Ukraine and Byelorussia. The city of Pripyat and the sarcophagus only form a small part of the vast contaminated area that Filatova calls the Land of Wolves. It is an exclusion zone in the Ukraine, and it is called the "Radiation Ecological Forest Reserve" in Byelorussia. According to Filatova (2005), new maps do not show the old roads, and places of human habitations, and this fact underscores the wasteland characteristic of the area. The area has a varied wildlife, wild boars, horses, and wolves. Vegetation is lush, engulfing the signs of former human presence. During evacuation cattle were slaughtered, but the offspring of cats live in the woods today (Filatova 2005). What is appealing about the exclusion zone is its material transcendence, the nuclear disaster pushing human beings away and allowing a glimpse into the underlying material world.

The Land of Wolves is also a memorial to human endeavor, visible in failure. Filatova sees the sarcophagus as modernity's lasting imprint on history: "The sarcophagus will remain radioactive for at least 100,000 years. The age for the pyramids of Egypt is 5,000 to 6,000 years. Each cultural epoch left something to humanity, something immortal, like Judaic epoch left us Bible, Greek culture – philosophy, Romans contributed law and we are leaving Sarcophagus, the construction that is going to outlive all other signs of our epoch and may last longer than the pyramids" (2005). How will the sarcophagus be viewed in the future? As a monument to ecological hubris? As a valiant effort to control nature that underestimated its material complexity? A monument to death, like the pyramids? Or a monument to creativity and risk-taking? Filatova is moved by the emptiness in a school: "There will be no more lessons in this school classroom, either. The only lesson taught here now is that the physical half-life of Americium-241, the daughter of Plutonium-241 is more than 400 years" (2005). Chernobyl teaches the lesson of possible large-scale incongruence between projection and materiality. How this incongruence is interpreted in a cultural context, however, is undetermined. The timelessness that Filatova describes means that the zone is removed from the human utilitarian clock and returned to geological time. In this sense the experience of the exclusion zone matches that of the desert, described by J. B. Jackson: "For what makes the landscape so impressive and beautiful is that it teaches no copybook moral, no ecological or social lesson. It simply tells us that there is another way of measuring time and that the present is, in fact, an enormous interval in which even the newest of man-made structures are contemporary with the primeval" (1994, 17).

The material transcendence of the zone evokes a paradoxical feeling of unreality, enhanced by its soundscape: "The silence is tremendous. No birds singing, no wind, nothing that can break this silence. Villages more picturesque than towns, houses and sheds do not look real. All look painted and I feel like I walk inside of this painting" (Filatova 2005). The sense of unreality is caused by the breakdown of projection, but there is no direct access to the material world. The underlying reality of radiation can only be perceived with a dosimeter; several photos in Filatova's essays show lush, pastoral greenery and a dosimeter in the foreground. A dosimeter is needed because radiation is distributed unevenly in the landscape. Aesthetics, usually limited to sensory perception, can incorporate scientific non-sensory data to make the invisible visible.

What is the lesson learned in the beautiful anti-landscape of the Land of Wolves? There is no fundamental difference between natural and human-made disasters, between whiteouts and blackouts. Both are limitations imposed by the material world. The anti-landscape is not a rejection of the human sphere, but a sign that human control over the world is limited. Limitation does not negate human endeavor but is its very essence, as there is no touch without material resistance. Throughout most of human history, the struggle with nature was the prerequisite for organization and solidarity. Only the blindness to material limitation has created an alienated hedonistic consumerism that forgot that civilization must be maintained in a struggle. Through industrialization the nature of this struggle has changed, as its main aim is not to subdue nature and to tame the whiteout but to avoid the negation of the material world, the blackout. This avoidance of negation and acceptance of limits could be the basis for a modern environmental awareness.

It might be argued that the anti-landscape suggests an updated epic sense of human endeavor. The epic theme has understandably been discarded by environmentalism (assuming that it is based on a struggle with nature with the aim to conquer it) and been supplanted by the notion of a harmonious nature into which humanity needs to integrate itself. However, both versions of human relation with nature, the heroic and the harmonious, are based on a false ideological premise, namely that nature is meaningful and predictable. A new epic sense of nature could stress the need to avoid negation through the material world as well as to accept that nature is wasteful and chaotic. Žižek argues that "'[t] he world without us' is thus fantasy at its purest: witnessing the Earth itself regaining its pre-castrated state of innocence, before we humans spoiled it in our hubris. The irony is that the most obvious example is the catastrophe at Chernobyl: flourishing nature has taken over the disintegrating debris of the nearby city of Pripyat, which had to be abandoned" (2011, 80). There is no pre-lapsarian state of affairs, and the new epic quality in the relation to nature could be seen in learning to live with unpredictability, not in domination or projecting harmony onto the world.

5. Suburbia: Anti-landscape of internal contradiction

The suburb appears to be an unlikely candidate for being an anti-landscape, and Robert Fishman's definition of it does not conjure up images of toxic dumps: "The suburb must be large enough and homogenous enough to form a distinctive low density environment defined by the primacy of the single family house set in the greenery of an open, parklike setting" (1987, 5). However, as the *Wall Street* journalist Lee Siegel notices, the suburb is the screen for negative projections: "For Yates, Plath, Ginsberg and less gifted suburb-phobes […], as well as hugely influential liberal sociologists and writers like David Riesman, William Whyte, Paul Goodman and Betty Friedan, it went without saying that the suburbs could transform the people who had committed the error of moving to them into the walking – make that driving – dead" (2008). Tuan makes a similar statement: "Among well-educated and sophisticated people, attitudes to suburban living are ambivalent: a professor of literature will admit to an address on the Greenacres with the same sheepishness that he admits to owning a color TV" (1990, 225). The suburb has an ambivalent role in Western (particularly American) society, embodying contradictory cultural desires. The reasons for the negative projection are complex: they originate both in an incongruence between material world and projections onto it and in an internal ideological contradiction in the projection itself. The suburb therefore is an anti-landscape that combines the hitherto discussed mechanisms of negation.

The limits of material projections with regard to the suburb are well known. The suburb is built on the projection of unlimited energy for cars and unlimited space for development. The current discussions about peak oil, rising food prices (caused partially by conversion of agricultural land), and the financial crisis that has been caused by real estate speculation are beginning to show the limits of that projection. There are also economic reasons; suburbs in the Rust Belt are Fourierite communities being abandoned by a society that consumes but does not produce. Searching the Internet for "abandoned" and "suburb," one finds many images of deserted areas and overgrown houses (for example, DETROITURBEX.COM 2011). Christopher Leinberger argues that the suburbs may be the "next slum" because of a "major shift in the way many Americans want to live and work" (2008). Demographic changes such as the increasing number of single households, the older "empty nesters," as well as energy costs and the convenience of urban living are the driving forces, accompanied by the vicious circles of poor schooling and crime and a trend celebrating "Sex in the City" and deploring the suburban "Desperate Housewives" (the underlying reason here is economic segregation, the flight from the poor). Like the Land of Wolves, the anti-landscapes of abandoned suburban housing can be aesthetically appealing, as the photographer James D. Griffioen (2012) shows. His website contains many photos of vegetation taking over abandoned suburban "feral houses," and the reason for the aesthetic appeal is the physical resistance of the material world, creating material transcendence (in the case of Detroit, economic forces accompany material ones). As in Filatova's photos, the wasteland creates an epic sense of failure without providing a clear moral lesson. However, the suburb is more complex than Chernobyl because it not only is a monument of failed material projections but also of internal ideological contradictions. The same cultural ideals that the suburb embodies are felt to be betrayed by it. Here I will consider the ideals of a sense of place, of personal relations, of future-orientation, and of individualism.

The suburb negates a sense of place. It promises a closer relation to the land and a sense of place through individual land ownership. The earthiness of one's own garden is, however, contradicted not only by environmental unsustainability. According to James Howard Kunstler, the focus on land ownership means "the sacrifice of a sense of place: the idea that people and things exist in some sort of continuity, that we belong to the world physically and chronologically, and that we know where we are" (1993, 118). But the land, including its climate and economic activities, does not through its settlement patterns and land use create a sense of place. For the suburb is a projection of consumerism onto the land. Consumption overrides the material limits of an area, and it overrides local economies built on local resources. According to Fishman "the picturesque landscape has been carefully designed to represent the consumption of the property by the viewer/owner, and this passive enjoyment is precisely the relationship of suburbia to its environment" (1987, 49). This consumptive attitude leads to a perceived lack of authenticity and undermines a sense of place.

The suburb negates personal relations. Fishman calls the suburb a bourgeois utopia, a "vision of the modern family freed of the corruption of the city, restored to harmony with nature, endowed with wealth and independence yet protected by a close-knit, stable community" (1987, x). One of its pillars is the separation of economic and private lives, a trend that started in London during industrialization. Suburbanites wanted to escape the moral temptations and corruptions of the city, to remove their families from the roughness of the economic world: "The bourgeois residence was now freed from traditional patterns to be redesigned as a wholly domestic environment – the home of a family that acted primarily as an emotional rather than an economic unit" (Fishman 1987, 9-10). Nevertheless, many writers present personal relations in the suburb as commodified. In Arthur Miller's *Death of a Salesman*, the family communicates in marketing speak; Linda advises Biff how to deal with his father:

> And have a nice time with Dad. He may have big news too! … That's right, a New York job. And be sweet to him tonight, dear. Be loving to him. Because he's only a little boat looking for a harbor. [*she is trembling with sorrow and joy.*] Oh, that's wonderful, Biff, you'll save his life. Thanks, darling. Just put your arms around him when he comes into the restaurant. Give him a smile. That's the boy … (2000, 59)

In Sam Mendes' film "American Beauty" Carolyn tells her husband to "act happy" when they go to a business reception, because the image of a good marriage is important for success. Betty Friedan describes the life of the suburban housewife as frustrating and lacking emotional fulfillment exactly because of the separation of the economic and the emotional spheres. What remains if both economic and emotional ties dissolve? In Mendes' film "Revolutionary Road" April reacts with "I don't feel anything" when her husband Frank confesses an affair, and Lester Burnham in "American Beauty" says that he feels "sedated." Karl Marx (1848) describes the effect of capitalist economic relations on personal ones: "The bourgeoisie has torn away from the family its sentimental veil, and has reduced the family relation to a mere money relation." The flight to the suburb accelerated this commodification of personal relations. Its lack of engagement in the (social and natural) environment that it embodies creates an effect of non-responsiveness, leaving the suburbanites, like Lem's space researchers or Tarkovsky's zone tourists, with mirrors, reflecting their own contradictory desires.

The suburb negates change and chance. Fishman points out that the suburb has to be understood in terms of its "rejected opposite," perceived as the "inhumane and immoral metropolis" (1987, 27). This flight was not an escape from the solvable problems of higher population density such as hygiene, but was a flight into a static world, away from modernity and its unpredictability. This sentiment is illustrated in the film "Revolutionary Road," where the dream of living an unpredictable life in Paris is pitched against the predictability of the suburb. The couple with the telling name "the Wheelers" gets stuck in suburbia when they trade in an exciting chance for material benefits. It is a paradox that "the economic system that created the suburb cannot provide stability and permanence" (Kunstler 1993, 170). "American Beauty" captures the dissonance between the timelessness of the suburb and a volatile job market that constantly evaluates "who is valuable and who is expendable." Rebecca Solnit discusses the design element of this suburban stasis: "The new architecture and urban design of segregation could be called Calvinist: they reflect a desire to live in a world of predestination rather than chance, to strip the world of its wide-open possibilities and replace them with freedom of choice in the marketplace" (Solnit 2001, 255). The timelessness of the suburb matches that of the Land of Wolves or of the abandoned suburbs of the Rust Belt. The suburb is both a utopia, escaping the upheavals of the industrial age, and a dystopia, generating a feeling of being frozen in time. An interesting moment is the transition from utopia to dystopia, seen for example in "American Beauty" where the utopian dream is pinched by material transcendence, the foreshadowing of doom; we see vegetation highly controlled in the film (e.g. obsessive pruning), but seeing it today, one cannot suppress images of trees breaking through floors in feral houses or in Pripyat.

The suburb negates individualism and privacy. Kunstler points out that there has been a discourse on conformity in the suburb since the 1950s (1993, 183), the fear of being a "dime a dozen." In "Revolutionary Road" there is a haunting image of office worker clones exiting the commuter train, and "A Serious Man" shows a horizon of sameness, of identical houses. The suburb has become a symbol for the fact that individualism in advanced capitalism is reduced to consumer choice. This includes the former cultural battleground of sexual identity. "American Beauty" ironically portrays a lesbian couple with identical haircuts and blurs the features of the gay neighbors of the Burnhams into exchangeable lawyer types, presenting themselves as "partners" and eliciting the question from their army neighbor: "What are you guys selling?" The disappearance of individualism has economic roots, as Michael Hardt points out: "[T]he production of the refrigerator and the automobile are only midpoints for the creation of the labor and gender relations of the nuclear family around the refrigerator and the mass society of individuals isolated together in their cars on the freeway" (2010, 142). Capitalism is based on individualism but also undermines it, as individuals are first transformed from producers into consumers and then, as for example by Google and Facebook, from consumers to the raw data material being sold to advertisers. Kunstler points out that the "extreme individualism of property ownership [...] tends to degrade the idea of the public realm" (1993, 26-27), and it is this lack of public space that paradoxically undermines privacy. Jackson argues that the suburb answers the "persistent need for privacy" through providing a "freestanding dwelling on its own well-defined plot of land" (1994, 157), but privacy conflicts with property, with the lack of public space. The suburb resembles a panopticon, and it is in this sense different from both the city and nature, where one can hide or be anonymous. The uncontrolled spaces in the suburb, however, are few and prone to invasion. In "American Beauty," Carolyn's affair ends when she is seen with her lover, Lester smokes a joint with Ricky but is soon found by his wife, and Ricky must keep his door unlocked. Ricky's camera, with which he records his neighbors, is a symbol for this pervasive visibility. The projection of freedom onto space has led to social control. This is not surprising because the idea of generating freedom through ownership is based on control of space, not on the openness of space.

Illustration 1. Richfield Minnesota, 1950. Richfield Historical Society / Wikimedia Commons

The suburb negates verticality. The reduction of social dimensions in the suburb is mirrored by two-dimensionality in real space. Horizontality is not the same as being close to the ground. The historical rootlessness of the suburb and its environmental unsustainability negate the earth, and its anti-modern attitude negates reaching out into the sky. It creates a feeling of simultaneous agoraphobia and claustrophobia, of two-dimensional infinity, of being stuck in endless space. Moving through this space on wide roads means "getting nowhere fast."

The suburb negates the past and the future. It appears to emulate small-town life but negates both traditional and modern lifestyles. Kunstler sees the prevalent DIY-culture as a negation of local economic relations and workmanship (1993, 171) and blames the modernist rejection of the past for the homogenized suburbs. But this view is too narrow: the suburb negates not only the past but also the future. The individualism of the suburb exclusively accommodates the nuclear family, excluding collective models of living found among some religious groups such as the Amish, the Amana Community, the Shakers, the Hutterites, or the

Jewish kibbutz. Suburbia is equally hostile to secular experimentation of the sort that first emerged in the nineteenth century at the Brook Farm, Fruitlands, and the Fourierite communities, and which re-emerged in the land communes of the 1960s. Suburbia can also be seen as a bastion against the open relationships advocated by the communist Sexpol movement of the interwar era, as well as the architectural avant-garde of the early Soviet years: the Byt-movement advocated communal living and began constructing buildings that facilitated communal eating and child rearing. Stalin soon denounced this movement as "leftist utopianism" (Hatherley, 2008, 54) and abandoned social experimentation. Both consumer capitalism and Stalinism negate collective alternatives, leading to what Owen Hatherley calls a "degeneration of the sexual-utopian imagery" (2008, 85).

A voice from a more future-oriented age is Ernest Callenbach's unabashedly utopian novel *Ecotopia*. Cities in Ecotopia have shed their suburban areas. The cover of the book illustrates the ideal of densely settled areas set in vast natural spaces. Likewise, public transport replaces automobile traffic in Ecotopia. Callenbach recycles the Aristotelian city ideal, arguing that settlements should be "microcities," consisting of some tens of thousands inhabitants, a short distance from a train station. He proposes to make economic activity the center of life, thus re-introducing a sense of engagement in the world, and to replace core families with collective living arrangements. Callenbach's novel is not a recipe for a pastoral return to a golden age but a bold imagination of creating a better collective future with modern means. Today the term "utopia" is negatively charged, referring to failed social experiments. However, we do not live in a post-utopian age, as Žižek points out, but the negation of utopia is itself utopian: "The true utopia is the belief that the existing global system can reproduce itself indefinitely; the only way to be truly 'realistic' is to think what, within the coordinates of this system, cannot but appear as impossible" (2011, 363). Today the "rejected opposite" of suburbia is not the gentrified industrial city but the blocks of collective housing found for example in the former Soviet Union. Many Westerners look at them with horror, the remnants of a failed dream, anti-landscapes of ideological meltdowns. But are they failures? The rejection of socialist housing is not so much based on its aesthetic uniformity but on a rejection of the underlying utopian vision of an egalitarian society. This is the Soviet version of the City upon a Hill.

Illustration 2. Murmansk, Russia 2010, photograph taken by the author

This image depicts part of a large housing area. Its castle-like vertical appearance appears intimidating but is an adaptation to the arctic climate. The harsh winds in Murmansk are not felt in the space inside the "castle" where children can play. There is also recreational space around the buildings, and the area is within walking distance of the city center and connected by public transportation. If one sees equality and community rather than individualism as guiding principles of housing, then the anti-landscape character of these blocks disappears. Also, those blocks are not a remnant of the past. New houses in Russia look similar to the old ones. The suburban mind is prone to see collective housing as an anti-landscape, but this gaze is reciprocated: from a collectivist utopian perspective, the suburb is an anti-landscape of capitalist "there is no alternative" fatalism. Even though there are Western-style suburbs in Russia today for the very rich, mostly the term suburb refers to substandard, often wooden, houses on the periphery for the disadvantaged classes, showing that cultural coordinates create the ideal of the suburb as well as anti-landscapes.

6. Conclusion

Landscape and anti-landscape are dialectical twins. Whereas landscapes can be protected, anti-landscapes are perceived as failures. The pastoral countryside, however, cannot be fully understood without the background of its rejected others, the desert or the dump. What is a landscape and what is an anti-landscape depends on the cultural lens of the viewer. Both landscape and anti-landscape have their origin in the interaction of the tangible material and the intangible cultural world. Anti-landscapes originate in cultural projection as well as in material transcendence, experienced as an unresponsive mirror or in the incongruence between projection and material reality. In addition there is the possibility that the projection of cultural meaning onto landscape is in itself incongruent or contradictory, and this is experienced by the visitors in Tarkovsky's zone as well as by suburbanites who feel that their projection of the ideas of individualism, freedom, nature, etc. are betrayed.

The anti-landscape highlights the "touching" aspect of landscape, the elements of limitation, negation, and failure that accompany human engagement in the world. This "touch" can be experienced with fear, as limit and failure, or with fascination, as material transcendence, suggesting a world behind projection. Since human engagement in the world is always prone to failure, anti-landscapes are common, and they are not moral predicaments but can engender a new epic sense of relating to the environment, of living with unpredictability. Material negation can never be overcome since the world is not ordered as our projections of it, but wasteful. Electricity and the sun do not make the night disappear; blackouts can suddenly switch off visual control, and also in a whiteout one experiences the disappearance of the horizon, confusing the sense of gravity. Both are disturbing because they remove control, but also exhilarating because they link us to the material world.

Bibliography

Abbey, Edward. 1991. *The Journey Home*. New York: Plume.

Bowles, Paul. 1990. *The Sheltering Sky*. New York: Vintage.

Callenbach, Ernest. 1990. *Ecotopia*. New York: Bantam Books.

Coen, Ethan and Joel Coen. 2009. "A Serious Man" (DVD). United Home Entertainment.

Cresswell, Tim. 2004. *Place: a short introduction*. Oxford: Blackwell Publishing.

Dawkins, Richard. 1998. *Unweaving the Rainbow: Science, Delusion and the Appetite for Wonder*. Boston: Houghton Mifflin Company.

DETROITURBEX.COM. 2011. "Abandoned Neighborhood in Highland Park." Available on-line at: http://detroiturbex.com/content/neighborhoods/hpdead/hpdead.html (last access April 4, 2011).

Elias, Norbert. 1986. "Über die Natur." Stuttgart: Merkur 40, 469-487.

Filatova, Elena. 2005. "Ghost Town and Land of the Wolves: Travelogue 2003-2005." Available on-line at: http://www.elenafilatova.com/ (last access March 19, 2011).

Fishman, Robert. 1987. *Bourgeois Utopias: The Rise and Fall of Suburbia*. N.p.: Basic Books.

Griffioen, James D. 2012. "Feral Houses" (photograph collection). Available on-line at: http://www.jamesgriffioen.net/index.php?/prairies/feral-houses/ (last access June 30, 2012).

Hardt, Michael. 2010. "The Common in Communism". In Douzinas, Costas and Slavoj Žižek (eds.). *The Idea of Communism*. London: Verso.

Hatherley, Owen. 2008. *Militant Modernism*. Winchester and Washington: Zero Books.

Jackson, John Brinckerhoff. 1994. *A Sense of Place, a Sense of Time*. New Haven: Yale UP.

King James Bible. 1769. Available on-line at: http://www.kingjamesbibleonline.org/ (last access March 5, 2011).

Kunstler, James Howard. 1993. *The Geography of Nowhere: The Rise and Decline of America's Man-Made Landscape*. New York: Touchstone.

Leinberger, Christopher B. 2008. "The Next Slum? The subprime crisis is just the tip of the iceberg. Fundamental changes in American life may turn today's McMansions into tomorrow's tenements." *The Atlantic*, March 2008. Available on-line at: http://www.theatlantic.com/magazine/print/2008/03/the-next-slum/6653/ (last access November 20, 2011).

Lem, Stanislaw. 2003. *Solaris*. London: Faber and Faber.

Marx, Karl. 1848. *Manifesto of the Communist Party*. Available on-line at: http://www.marxists.org/archive/marx/works/1848/communist-manifesto/ch01.htm#007 (last access April 2, 2011).

Melville, Herman. 1856. "The Encantadas or Enchanted Islands." Available on-line at: http://www.melville.org/encant.htm (last access March 5, 2011).

Mendes, Sam (director). 1999. "American Beauty" (DVD). Dreamwork Productions.

Mendes, Sam. 2008. "Revolutionary Road" (DVD). Dreamwork Productions.

Miller, Arthur. 2000. *Death of a Salesman*. London: Penguin.

Nye, David E. 2007. "Are Blackouts Landscapes?" *American Studies in Scandinavia*, 39:2, 2007, pp. 72-84.

Pinker, Stephen. 1997. *How the Mind Works*. New York: Norton.
Richfield Historical Society. 1950. "Photo showing Suburban development on Washburn Avenue, Richfield, Minnesota." Original photo in the collections of the Richfield Historical Society. Available on-line at: http://commons.wikimedia.org/wiki/File:Richfield_MN-washburn_Ave.jpg (last access November 23, 2011).
Siegel, Lee. 2008. "Why Does Hollywood Hate the Suburbs? America's long artistic tradition of claiming spiritual death by station wagon" in *The Wall Street Journal,* December 27, 2008. Available on-line at: http://online.wsj.com/article/SB123033369595836301.html (last access March 30, 2011).
Solnit, Rebecca. 2001. *Wanderlust: A History of Walking*. London: Verso.
Tarkovsky, Andrei (director). 2002. "Stalker" (DVD). Artificial Eye.
Tuan, Yi-Fu. 1990. *Topophilia: A Study of Environmental Perception, Attitudes, and Values*. New York: Columbia University Press.
Žižek, Slavoj. 2011. *Living in the End Times*. London: Verso.

9

Lost Landscapes: Degraded Landscape as Anti-Landscape

Maunu Häyrynen

This essay examines sites that achieved and then lost cultural importance, after which they either fell into oblivion or emerged with new meanings in an afterlife where they achieved a different significance. A range of theories are considered to deal with this phenomenon, followed by specific examples taken from Finnish history as illustrative case studies.

Keywords: Bakhtin, Mikhail; Danto, Arthur; Finland; Goffman, Erving; Harvey, David; landscape; Mitchell, W.T.J.; nostalgia; national identity; postmodernism

1. Grasping anti-landscape

'Anti-landscape' seems intuitively a straightforward thing: Something against landscape. If landscape is seen as beautiful, anti-landscape is ugly; if natural, anti-landscape is urban; if useful, anti-landscape is useless; if harmonious, anti-landscape must then be chaotic. Anti-landscape thus understood is relational, each kind of landscape representation generating its own antithesis. Anti-landscapes are landscapes all the same, however gruesome or revolting – they merely mark the opposite ends of the generic and narrative register (Mitchell 1994a, 1).

Another way of grasping anti-landscape is to contrast represented landscape with what is left outside the frame: the obscured and eclipsed, such as the rural poor in John Barrell's *The Dark Side of Landscape* (1992), or the female or racial Other. One seems here to be dealing with non-landscape rather than anti-landscape, though. The logic of representation and the way of seeing based on it determine what is deemed worth depicting, imageable or picturesque, and what is ignored or marginalized as non-landscape.

A non-landscape might be pristine, an area not yet exposed to Western semiosis, should there be any. Otherwise it may be 'backstage', using Erving Goffman's term (1963), an intermediate space bracketed off from representation. Backstage too has the potential of being represented: the boundary line between landscape and non-landscape is ill defined and oscillating. Indeed, following Jean Baudrillard's or Guy Debord's (1995) criticisms, there is no longer any boundary line but only endless circular signification.

In the Marxist tradition landscape representation is discussed in terms of power and appropriation, expressed in individual terms and linking with the ruling classes and the production logic of capitalism. Visual landscape has provided a means of spatial occupation and homogenization, leading to the global hegemony of the Western concept of landscape. David Harvey (1995) has described this process as "time-space compression", along which place experience becomes increasingly replaced by mediated place imagery. The economic competition of localities becomes expressed as ever expanding, accumulating and competitive imagery. Harvey sees a fundamental contradiction between the fixity of place and the spatial mobility of capital, in which landscapes shaped during one phase become barriers of development in the next one. In this process old places are devalued, destroyed and redeveloped into new ones (Harvey 1996).

The dynamic works both ways, so that in addition to what is yet to be represented one may speak of what has ceased to be represented, places that have been held as landscapes but no longer are so. The difference is gradual rather than absolute. Landscape imagery works as a kind of abstracted meta-'glance' (Nuti 1994), highlighting certain localities and overshadowing others. Non-landscapes in this respect are the ones that somehow fail to conform to the prevailing hegemony (Cosgrove 1989).

Anti-landscape appears in this respect a strategy of resistance and decentering, occurring as counter-imagery or deconstructing the representational logic by deliberate misuse of pictorial conventions and redirection of ideological connotations. Art offers a wealth of examples of distorting or displacing, including commercial messages, that overthrow existing scopic regimes of landscape, disclosing their underlying power relations and suggesting alternative ways of seeing.

Here a closer comparison between art and landscape might prove useful. Arthur Danto (1964) introduced the concept of 'anti-art' as an indispensable part of the dynamics of what he called artworld, whereas non-art would not be deemed as art unless sanctioned by the artworld. Without probing further into this discussion, anti-art may be likened to intentional contestation of landscape, whereas the backstage would correspond to non-art – remaining non-landscape until the moment of representation. The boundaries of the art world are easily transgressed but anti-art remains closer to art than non-art, forming a part of the same discourse. In a similar vein the deconstruction strategies aimed at landscape (see Debord 1995) have ended up adding a new layer of meaning to it.

W. J. T. Mitchell sees the semiotic and hermeneutic view of landscape as an allegorical text that is problematic and reductionist. He presents an alternative concept of an autonomous and active landscape medium that does not merely signify or symbolize (or contest) power relations. Landscape according to him is a relatively independent agent of power not retraceable to human intentions but still shaping both real and represented environments. His question is, what landscape as a body of cultural and economic practices *does* rather than what landscape 'is' or 'means' (Mitchell 1994a, 1-2, his italics; Mitchell 1994b, 5). In Bruno Latour's terms the landscape medium is a hybrid, reducible neither to human nor natural agency (Latour 1993).

Mitchell's landscape medium articulates the relations not only between social groups but also between man and nature. Mitchell himself posits as the opposite of landscape 'land' or, more precisely, the "violence and evil written on the land – by the gazing eye" of imperialism and nationalism and consequently masked and naturalized by the landscape medium (Mitchell 1994b, 15 & 29-30). This contrasting of representation and the represented echoes Raymond Williams' (1973) earlier distinction between landscape and the working countryside.

Mikhail Bakhtin's dialogical landscape is an attempt to build a bridge between representation and historically evolving community. Landscapes in his sense can be regarded as "repositories of polyphony and heteroglossia: places where social, historical, and geographical conditions allow different voices to express themselves differently than they would under any other conditions" (Folch-Serra 1990, 256). These are at the same time communicative and polyphonic landscapes and chronotopes, in which historical time becomes visible in physical space. The dialogical landscape has something in common with the phenomenological understandings of the place as meaningful, experiential and changing over time, its opposite being the bleak and ubiquitous blandscape of globalization (Cresswell 2004, 19-33). There are also other approaches defining landscape as a specific social order rather than representation (Olwig 2002; Jones 2008). The 'anti-landscape' would, in this respect, appear as a blank slate, sustaining no community, dialogue or social practices.

2. Anti-landscape and landscape imagery

At least four dimensions of anti-landscape have been sketched above: the textual anti-landscapes, the redundant landscapes in the way of progress, the counter-landscapes of contestation and the monotonous blandscape. In the following I try to locate aspects of anti-landscape in the historical trajectories of Finnish landscapes, based on my previous study of popular nationalistic landscape imagery.

By nationalistic landscape imagery I mean a representational system based on national territory and its occurrence in popular media. Such imagery is present in the production of imagined geographies and spatial hierarchies, naturalizing national space and anchoring it in concrete lived environments. Landscape imagery produces a more or less fixed set of core images that are repeatedly represented in different contexts. Core images turn over time into undisputed landscape 'classics' (Häyrynen 2008 e.g.) associated with actual places, as it were carving them out of everyday lived space. Similar notions about imagined national space anchored on existing physical sites have developed by others, notably Rob Shields (1991), Pierre Nora (1996) and Sverker Sörlin (1998).

The verisimilitude or indexical power of landscape imagery rests on the possibility of personally experiencing the places that are depicted. The bodily experience of the sites regenerates the hierarchy and ideological connotations of landscape imagery. A self-enforcing circle is formed in which imagery fuels tourism that in turn promotes landscape classics. Landscape depiction is accompanied with a range of other visual/spatial technologies, which entail the opening and manipulation of views, the building of lookout points, the screening out of unwanted elements and so forth. These technologies have different genealogies – mapmaking and surveying, fortification and reconnaissance, architecture and garden art – but have become interconnected with the formation of landscape imagery. In the repeatedly represented places one visualization technology or another has usually been applied to their physical settings in order to preserve and enhance their picturesque qualities. In certain spots visuality may even have conditioned other land uses, turning them into specifically staged landscapes.

The status of being an iconic landscape detaches a place from the surrounding everyday environment. This is a familiar pattern in touristic landscapes. There are alternative development paths or trajectories the interaction between the site and its representation may follow, passing through phases such as initial recognition, establishment, fossilization or reshaping and forgetting or renewed interest. There are a number of drivers behind the different trajectories. A major turn in the aesthetics and mode of representation of landscape, such as the eighteenth century picturesque turn in England (Andrews 1989), changes the conventions for appreciating and representing landscapes. Profound changes in land uses alter – sometimes with a time lapse – the way a place is scenically valued and framed. Also the accessibility of a site may vary. Last but not least, the geopolitics and the social power relations behind the landscape may be replaced.

What seems to be important for the long-term trajectory of landscape is its multicontextuality, its articulation in several contexts and by different media and institutions simultaneously (Sörlin 1998). Landscape classics are not unlike artworks as auratic objects (Benjamin 2005). They are constructed gradually and gain additional strength from the inclusion of the place in new domains: from science to art, from art to administration and so forth. The more established a landscape, the more likely it is to become institutionally protected and maintained, even staged.

Repeated representation of a landscape leads to its banalization as cliché and loss of uniqueness or aura (Wilson 2005). Nevertheless, some landscapes continue being represented in changing contexts while others may be forgotten. What exactly decides between canonization and degradation is often hard to identify, but the degrading processes are interesting in themselves. Banalization often follows both popularity and development. A common example is the flight of the elite from congested tourist resorts that have lost their one-time authenticity and distinctiveness (see e.g. MacCannell 1989).

The process of degradation is in a way symmetrical to the emergence of a classic. It begins with the end of tourism and is followed by a site's gradual disappearance from popular media. Scenic values become disregarded in land use or the maintenance of the site stops altogether. Tourism infrastructure and staging elements like vista lines and screens, lookout places or scenic routes, lose their meaning and erode. The landscape may later be resurrected as a place of memory and an object for nostalgia, starting a new cycle in its trajectory.

3. Trajectories: Landscape to anti-landscape

Representation of Finnish landscape was begun by foreign travellers, artists and scholars in the late eighteenth century. For them the entire geographic area of Finland was interesting as a Sublime periphery. Sublime originally was the generic opposite to pastoral beauty, contrasting wild natural scenery with the tamed cultivated landscape. Among the key motifs especially in South Finland were the then free-flowing rapids, typical for a country with a plenty of watercourses and small-scale height variation. The most spectacular among them would compare with well-known benchmarks such as Schaffenhausen, Trollhättan or

Illustration 1: The Imatra Rapids were the first and foremost natural sight in Finland prior to their harnessing in 1930. Litograph by Erik Kruskopf in Zachris Topelius: Finland framstäldt i teckningar, Borgå 1845.

even Niagara, and could be captured without difficulties by the existing pictorial conventions. As a consequence they became widely mediated by popular *voyage pittoresque* illustrated travel accounts and started to attract elite tourists from abroad. These were followed by Finnish artists, translating the international Sublime into patriotic symbols.

This situation was profoundly changed by the relatively late industrialization of Finland, accelerating in early twentieth century. The rapids were harnessed for hydropower production one after another, following bitter disputes between the proponents of traditional national Romanticism and advocates of economic progress. What had started as an emerging alternative landscape more than a century earlier was later perceived as a redundant landscape that could be pushed aside by the onslaught of capitalism. There were attempts to convert some of the harnessed rapids into monumental Modernist landscapes by means of architectural and landscape design. Despite these efforts, they practically disappeared from popular landscape imagery, having lost their scenic and touristic appeal.

Illustration 2: Finnish National Romanticist painter Eero Järnefelt standing on top of the Koli heights in 1936. Finnish National Board of Antiquities.

The multicontextual landscape of the foremost rapids was literally drowned under reservoirs and replaced by monotonous techno-landscape. Only one of them, the former rapids of Imatra, has been reframed as a national iconic landscape on the basis of representational history and tourism.

Another case is provided by a group of panoramic landscapes, located near the Russian border that were depicted by National Romantic artists in the late nineteenth century (Häyrynen 2004). The "Golden Age" landscape painting created a panoramic landscape type considered quintessentially Finnish, a boundless wilderness extending to the feet of the spectator. This image was repeatedly mediated by photographic landscape imagery through popular printed media. Some of the sites became accessible by modern transportation and their touristic development was initiated, to be followed by first conservation measures in the early twentieth century. They normally consisted of a hotel in a historicizing style, a system of pathways, lookout places and vistas as well as a protected and managed forest screen maintaining an illusion of wilderness. The later development of panoramic landscapes took different paths. One of them, Vuokatti, fell prey to market forces, turning the site into a winter sport resort that was redeveloped without any further regard to landscape. Another one, Koli, was expected to share this fate in the 1980s, but due to a massive public protest movement it was instead designated as a National Park in 1991.

Illustration 3: Modern-day three-dimensional ski slope map of the Vuokatti heights. Huippupaikat Oy, Vuokatinrinteet.

The third site, Riuttavuori, was left behind the new border with the Soviet Union after the World War II. The entire area in effect disappeared so far as the Finns were concerned as a new population displaced the former Finnish residents who migrated to the remaining part of their country. The new inhabitants were unaware of the symbolic status or the history of representation of the site, which also includes a prehistoric hillfort. The Soviet and later Russian landscape representations of the area focused on built urban and vernacular landscape and on generic images of nature, and did not share the Finnish penchant for panoramic wilderness landscapes. As a result of this dramatic recontextualization process, the site of Riuttavuori was completely abandoned and ignored by the local community, and on the Finnish side of the border its significance was only sporadically maintained in a nostalgic rememberance discourse.

To conclude, the above examples of landscape trajectories exemplify several dimensions of anti-landscape. Both the technolandscapes of the harnessed rapids and the touristic landscapes of earlier wilderness landscapes are anti-landscapes in the sense of David Harvey, places made redundant, redeveloped and destroyed by the swift movements of capital. Their specificity, produced by a polyphonic and multicontextual interaction and contestation process, has been replaced by anonymous blandscape, another aspect of anti-landscape. Lastly, Riuttavuori is the ultimate case of anti-landscape in which landscape has been stripped of community relations, social practices and dialogic exchange, only leaving the discursive landscape of memory.

Bibliography

Andrews, Malcolm. 1989. *The Search for the Picturesque. Landscape Aesthetics and Tourism in Britain*. Aldershot: Scolar Press.

Benjamin, Walter. 2005 (1936). *The Work of Art in the Age of Mechanical Reproduction*. UCLA., http://www.marxists.org/reference/subject/philosophy/works/ge/benjamin.htm.

Cosgrove, Denis. 1989. Geography is Everywhere: Culture and symbolism in human landscapes. In: *Horizons in Human Geography*. Gregory, D. & Walford, R. (eds.). Houndsmills: MacMillan.

Cresswell, Tim. 2004. *Place, a short introduction*. Cambridge: Blackwells.

Danto, Arthur C. 1964. The Artworld. *Journal of Philosophy* 61 (1964), 571-584.

Debord, Guy. 1995 (1967). *The Society of the Spectacle*. Translated by D. Nicholson-Smith. New York: Zone Books.

Facos, Michelle. 1998. *Nationalism and the Nordic Imagination. Swedish Art of the 1890s.* Berkeley: University of California Press.

Folch-Serra, M. 1990. "Place, voice, space: Mikhail Bakhtin's dialogical landscapes." *Environment and Planning D: Society and Space* 8 (1990), 255-274.

Goffman, Erving. 1963. *Behavior in Public Places. Notes on the social organization of gatherings.* New York.

Harvey, David. 1995 (1989). *The Condition of Postmodernity. An enquiry into the origins of cultural change.* Cambridge: Blackwell.

Harvey, David. 1996. *Justice, Nature and the Geography of Difference.* Cambridge: Blackwell.

Häyrynen, Maunu. 2004. "A periphery lost: the representation of Karelia in Finnish national landscape imagery." *Fennia* 182: 1 (2004), 23-32.

Häyrynen, Maunu. 2008. "A Kaleidoscopic Nation: The Finnish National Landscape Imagery." In Jones, Michael & Olwig, Kenneth (eds.): *Nordic Landscapes, Region and Belonging on the Northern Edge of Europe.* Minneapolis: University of Minnesota Press.

Jones, Michael. 2008. "Landscape – morphology or community? The role of land rights, customs and local institutions for shaping the land in Nordic agrarian societies." In Jones, Michael, & Olwig, Kenneth (eds.): *Nordic Landscapes, Region and Belonging on the Northern Edge of Europe.* Minneapolis: University of Minnesota Press.

Latour, Bruno. 1993 (1991). *We have never been modern. Transl. Catherine Porter. Cambridge: Harvard University Press.*

MacCannell, Dean. 1989. *The Tourist: a new theory of the leisure class.* Schocken Books, New York.

Massey, Doreen. 1997. "A Global Sense of Place." In Barnes, T. & Gregory, D., eds. *Reading Human Geography.* London: Arnold, 315-323.

Mitchell, W. J. T. 1994a. "Introduction." In Mitchell, W. J. T., ed. *Landscape and Power.* Chicago: University of Chicago Press.

Mitchell, W. J. T. 1994b. "Imperial Landscape". In Mitchell, W. J. T., ed. *Landscape and Power.* Chicago: University of Chicago Press.

Nora, Pierre. 1996 (1984-1992), ed. *Realms of Memory*, I–III. New York: Columbia University Press.

Nuti, Lucia. 1994. The Perspective Plan in the 16th Century: The invention of representational language. *The Art Bulletin* LXXVI (1), 105-128.

Olwig, Kenneth. 2002. *Landscape, Nature and the Body Politic.* Madison: University of Wisconsin Press.

Shields, Rob. 1991. *Places on the Margin.* London: Routledge.

Sörlin, Sverker. 1998. Monument and Memory: Landscape imagery and the articulation of territory. In *Worldviews: Environment, Culture, Religion*, 2, 269-279.

William, Raymond 1973. *The Country and the City.* London: Chatto & Windus.

Wilson, Paul 2005. "Banality as Critique: Contemporary Photography and Finnish National Landscapes: (Contested) Finnish National Landscapes." *Picture, Place, and Power Symposium*, Minneapolis, Minnesota 24 September 2005.

III

Recoveries

10

Landscapes of Waste: Malmberget and Ignalina as Cultural Tools in Heritage Processes

Anna Storm

This essay considers two industrial sites, an old open-pit mine in Malmberget in northern Sweden and the closed Ignalina nuclear power plant in northeastern Lithuania, to see to what extent each might possibly become part of local heritage processes. Both sites once represented visions of progress and modernity, but today they present problems of waste and unemployment that evoke ambiguous feelings in their communities. These are wounded landscapes that are difficult to heal. However, the physical scars are also part of the stories that are the fabric of heritage, and these landscapes of waste may reemerge as new landscapes of memory and community identity that are directed toward the future.

Keywords: heritage process; Ignalina; industrial landscape; Lithuania; Malmberget; mining; nuclear power; Sweden; waste

1. Introduction

A huge mining pit in the midst of Malmberget in northern Sweden and contaminated structures and radioactive spent fuel at Ignalina nuclear power plant in northeastern Lithuania constitute not only urgent material problems to be dealt with but also potential cultural tools in local heritage processes. The challenges of handling the waste and other physical traces of industrial production will in the following be examined in relation to stories about the significance of these landscapes. Would it be possible to activate their complex and ambiguous connotations within a process of articulating meaning and value in the present situation?

This article is based on original research in the mining town of Malmberget in Sweden as well as the Ignalina nuclear power plant and the connected workers' town of Visaginas in Lithuania, carried out in 2007 and 2010 respectively (Olsson and Storm 2008; Storm 2010). Inspired by the concept of anti-landscapes, I found it productive to combine these two sites in order to elucidate a specific perspective. On the one hand, what made these sites appear as repulsive, dangerous and useless, and, on the other hand, how do they support meaning and value in the local communities? If anti-landscapes primarily refer to spaces that literally do not sustain life as suggested in the introduction to this volume, this article will add a possible understanding of anti-landscape as referring to a landscape without articulated value.

Heritage is today generally understood as a cultural process "that engages with acts of remembering that work to create ways to understand and engage with the present" (Smith 2006, 44). It is often emphasised that it is meaning that gives value, while, for example, artefacts and sites in themselves have little intrinsic worth (see for example Graham and Howard 2008, 2; Gibson and Pendlebury 2009, 1). Nevertheless, a site or landscape can work as a tool or mnemonic device facilitating the heritage process (Smith 2006, 44: Gilloch 1997, 76) and it is in this respect the landscapes of industrial waste become subjects of investigation.

Until recently, heritage processes were often understood as something positive by definition, connected to the creation of knowledge and identity as well as experiences of beauty, and providing orientation in times of rapid change. However, during the last decades, more difficult heritage processes are increasingly being examined. These processes could be contested, ambiguous or unquestionably problematic, for example in relation to experiences of war and memories of abuses. One possible outcome of these complicated processes is that the difficulties of heritage articulation are somewhat defused or framed within contemporary progress, or even marked by reconciliation (see for example Macdonald 2009; Logan and Reeves eds. 2009; Jönsson and Svensson eds. 2005).

In a decisive way, the mining pit in Malmberget and the structures of the Ignalina nuclear power plant support meaning and value in their respective local communities. Nevertheless, the stories about their impact are neither articulated locally nor recognised within a wider geographical context. In other words, two potential key mnemonic devices are not activated as facilitators in cultural heritage processes.

In these two communities, characterised by a liminal status, decline and uncertain future prospects, such a process might presumably be important in order to engage with the present situation. What are then the stories that could be told, related to these two landscapes of industrial waste?

2. The Pit in Malmberget

Malmberget is a century-old mining town in the far north of Sweden, surrounded by forests and mountains.[1] Today it is a twin town to adjacent Gällivare, which emerged as a trading place in connection to the railway. Malmberget had its highest number of inhabitants in the 1960s, 12,000 people, out of which 3,000 were employed by the mining company LKAB (Luossavaara-Kiirunavaara AB). Today the town has half as many inhabitants and the mine employs 1,000 of them.

Among other things, the mining operations have resulted in a 200 meter deep pit, comprising an area of 21 hectares in the very heart of the town. The area around the pit is fenced off due to the risk of landslides. Initially it was an opencast mine that later grew considerably because of explosions and cave-ins, and it has affected the Malmberget community since the 1960s. Streets and building blocks have literally disappeared into the ground, and many significant public buildings have had to be demolished or moved. For example, in the beginning of the 1970s the church was moved to another part of the town, whereas the old school and the public assembly hall were demolished. The experience of adapting to the needs of mining activities has been prevalent in Malmberget for a long time. One can say that the mine simultaneously feeds and eats the town.

1. The description of the pit in Malmberget is based on Olsson and Storm 2008.

Illustration 1: A 200 meter deep pit dominates the town of Malmberget. Its continuous growth entails new houses and areas to be abandoned every year. Photo: LKAB.

Malmberget was a truly prosperous place in the mid-twentieth century, visible for example in its elaborate wooden houses, three cinemas, two swimming halls, an ice skating arena and a prominent high rise building of the sort not always found in a town of this size. Today the decline is visible. Although, for example, the former managing director's villa is not yet within the fenced zone around the pit, it is close enough to frighten away any potential buyers. The result is a building in an obviously dilapidated state. Also, a closer look at the town centre reveals that this place was once different. Empty shops and slightly overgrown public spaces give the centre a deserted character. The fluctuating world market for iron ore makes future prospects most uncertain, and the local community, for so long intimately intertwined with the mining company, now tries to establish new ways to build a stable economy, for example through tourism and specialty food production. But still the mine is the backbone for the local economy.

The local government commissioner in Malmberget describes the pit as something really worth seeing, something he shows important visitors, preferably from a helicopter, but also by going down into it by car. At

Illustration 2: The former managing director's villa, located close to the fenced zone around the pit, is falling into decay. Photo: Krister Olsson, 2007.

the same time, he notes that there is nothing good about the pit, displaying an ambiguous attitude that one also hears in statements by mining company representatives. Many people have grown up in the presence of the pit, and it is, hence, essential when it comes to understanding the post-war decades in the history of Malmberget, on an individual as well as a collective level. At the same time, however, the immediate area around the pit is covered by vegetation which makes it almost invisible from a ground level perspective. All that you see is young trees and roads that end unexpectedly in the distance.

The narrative about the town centre that disappeared into the pit, or had to be moved because of it, is strong and lively in Malmberget today. Nevertheless, in a long-term perspective, this will probably change. With coming generation shifts, one can expect that the story of disappeared important buildings and lost places of childhood will fade away, while other stories will take its place. As interviews with young people in Malmberget suggest, it is likely that narratives about the pit itself will come forward, that is, stories about fences and cave-ins, about border transgressions and risky situations connected for example to biking or climbing on the forbidden other side (Olshammar 2008). There is even a local saying: "the pit is coming" clearly illustrating that the pit is seen as almost a living being – it is still growing – moving closer to one's home and still accessible places (Feldmann 2008).

A few years ago, the mining company started to fill the pit, since it needed a waste rock dump, and in order to prevent cave-ins, although the main formal reason was to restore the landscape according to what is prescribed by legislation. At a pace of eight truckloads an hour, day and night, the filling was estimated to take ten years. The company did not hold out the prospect of recreating a land that it would be possible to build upon, but only the probability of using it as an area for walking and biking. In fact, company representatives asserted that the pit might never be filled completely because of the continuous mining activity underground. Apparently, no one really knows if it will ever be filled, or instead just shrink at one end and expand in another. Thus, the pit continues to be the pit, even though its shape and how it affects the town structure changes over time.

In a dramatic way, the pit in Malmberget represents an important past, of mining industry and welfare society, among other things, both on a local and on a national level. The local government commissioner declares that the pit is the only thing that makes the town special, but at the same time, no one in the local community or at the regional or national heritage authorities expresses any consideration about its potential value for telling this story. The pit constitutes a potential cultural tool that is not activated. The narratives connected to it show its meaning and value to different generations in the community, but the lack of articulation implies that a heritage process is not taking place. The landscape of the pit in Malmberget is still an anti-landscape in this respect, without explicitly stated value and with a meaning primarily connected to its role as a physical obstacle.

The mining industry makes up the very reason for the existence of the town of Malmberget, both in the past and in the present. Also in the town of Visaginas in Lithuania, the industrial activity, in this case nuclear power production at the Ignalina plant, was the basic reason for its existence. While mining is still going on at a small scale in Malmberget, the Ignalina plant shut down in 2009, although decommissioning and waste storage work still gives employment to a considerable number of people. Are there any similarities to be found between these two places? Or are the contrasts too big, due to the specifics of their varying industrial activities and disparate historical backgrounds?

3. The Ignalina Nuclear Power Plant

The Ignalina nuclear power plant and the accompanying town of Visaginas are located in the northeast corner of Lithuania, close to the borders to Latvia and Belarus.[2] When the plant and the town were established in the 1970s and 1980s under Soviet rule, those borders were of course of less importance. Visaginas, by many considered a socialist paradise (Cinis, Drémaité and Kalm 2008; Šliavaité 2005), was built on a previously "empty" spot in an area of forests and lakes, 150 kilometers from Vilnius. The number of inhabitants was about 25,000, out of which about 3,500 worked at the plant. Highly skilled engineers, relatively well paid, moved from other parts of the Soviet Union and only a few had a Lithuanian background. The town was a so called semi-closed town and thus given certain privileges, sometimes illustrated by the expression of "chocolate cites" because it was possibile to buy chocolate and other exclusive products there.

Today Ignalina's two reactors are shut down, due to demands from the EU. Nevertheless, about 1000 people are working at the plant, monitoring the radioactive material and preparing the structures to be torn down in one or two decades.

2. The description of Ignalina nuclear power plant and Visaginas is based on Storm 2010.

Illustration 3: Ignalina nuclear power plant with its two reactors and characteristic chimneys. The plan was to build four reactors at the site, but this was eventually not realised. Photo: Anna Storm, 2010.

The Ignalina nuclear power plant has made a remarkable symbolic trip from its inauguration until today. Being the world's biggest nuclear power plant when it was built and an emblem of the socialist frontier that employed Soviet heroes, it simultaneously became for Lithuanians a despised sign of Moscow influence during the Soviet period. Before an independence movement was formed, it was at times permitted to be an environmental activist, and so there was in the last years of the 1980s a lively environmental movement in Lithuania, with the main aim to close down Ignalina (Dawson 1996).

Shortly afterwards, when the Lithuanian struggle for independence gained momentum, the environmental movement was quickly transformed into an independence movement (Dawson 1996). Nevertheless, when independence was achieved in 1991, Ignalina – which provided Lithuania with 70 percent of its electricity – turned into the most valuable energy resource for the new independent nation. When Lithuania aspired to become a EU-member, however, the EU expressed severe concerns about any Chernobyl-type of nuclear reactors within its borders. Lithuanian EU membership was conditioned on closing down Ignalina. Consequently, the first reactor was closed down when the country entered the EU in 2004, and the second one in December 2009. In Visaginas today, there is an ongoing search for new ways to make a living (Šliavaitė 2005), although the closed plant and the deconstruction management, as mentioned above, still generate quite substantial work opportunities.

At the plant and in the town of Visaginas, there are few signs of the big industry being closed down. For example, the safety arrangements remain rigorous for anyone entering the plant and all spent fuel is still left in the reactor or in a water basin next to it, calling for 24 hours super-

Illustration 4: Ignalina nuclear power plant. A walk on top of one of the reactors, which still contains highly radioactive fuel. Photo: Anna Storm, 2010.

vision. And in the town centre, an electric sign placed under a sculpture of the symbol of Visaginas, a crane bird, still informs the inhabitants about the current temperature, time, and level of radiation.

In order to tear the nuclear power plant down, the waste has to be dealt with, and a huge storage building is currently under construction in the vicinity. According to the former managing director, the storage facility will last for 50 years and after that the radioactive waste will be a problem for the next generation to handle. This way of pushing the problems of waste ahead to future generations is not unique to Ignalina. In the world today, there are between 400 and 500 reactors and only two countries (Finland and Sweden) have final storage sites (built to last 100,000 years) in the process of being built.

To many in the older generations in Visaginas, Ignalina represents their whole working life, since many of them arrived in their twenties to work at the plant. Until its closure, the memories are predominantly bright and proud. People remember living in a beautiful place with everything they needed and being highly respected workers of the Soviet Union (Šliavaitė 2005). These are primarily native-speakers of Russian who experienced what they conceive of as betrayal through the Soviet collapse. Suddenly, they found themselves in the northeast corner of Lithuania, a country that needed their skill and the energy produced, but where a Russian-speaking minority was not well regarded (Šliavaitė 2005). For the ethnic Lithuanians, Ignalina was once the dark symbol of Moscow power and the focus of an independence movement in disguise. Later, it became a symbol of independent energy supply and finally a sacrifice to the EU demands in exchange for membership. A fear of possible nuclear accidents or of risks connected to the radioactive waste seems to be less pronounced in Lithuania than in some other nuclear nations.

Thus, as in the case of the pit in Malmberget, Ignalina represents an unquestionably important period in the life of the local community as well as an important feature of industry and politics on the national and international levels. Yet neither of these sites is considered important for its potential to support memory and to facilitate heritage processes. Instead, the goal is to make the pit and the plant disappear as completely as possible. Why is it so? Is it because industrial activity is still going on, at least in some sense, and a plan to end this activity is related to a responsibility to restore a kind of "natural" environment? Is it because of the dangerous and difficult aspects they reveal which not easily are interpreted as relevant in a cultural heritage process? Or is it the physical traces of industrial production that in themselves are deemed dangerous and useless and therefore block the ability to think in terms of heritage?

4. Difficult Heritage Processes

To contextualise the pit in Malmberget and the Ignalina nuclear power plant as potential cultural tools in heritage processes, a few examples of other former industrial sites are illuminating. Some have been abandoned while others have been turned into neat and tidy places. They often carry ambiguous significance, and they are often perceived as exciting or appreciated visually by a privileged audience who have no personal work experience in industrial environments. Concepts like "ruin tourism", "horror tourism" and "urban exploration" also pinpoint the wealthy world's fascination with difficult and dangerous heritage (see for example Isacson 2004; Ninjalicious 2005).

Among almost endless possible examples, I would like to mention the former Bankside power station in central London. For many years it stood empty on the southern shore of Thames River, facing the St Paul's Cathedral. Today it houses a famous art museum, the Tate Modern, and has become a tourist magnet. Another intriguing example is a former blast furnace in Duisburg, in the Ruhr area. This so called Meidericher Hütte was shut down in the 1980s and today the whole industrial area is transformed into what is termed a "landscape park" ("Landshaftspark") with facilities for exhibitions, walking, climbing, diving, biking et cetera (Storm 2008).

Illustration 5: Today, the former Meidericher Hütte in Duisburg in the Ruhr area is turned into a so called "landscape park". Among other things, the local mountaineering association uses the ore bunkers for climbing. Photo: Hans-Jürgen Wiese, 2006.

Sweden provides another example, in Gustavsberg, close to Stockholm, where former porcelain factory buildings are used as small shops, apartments and cafés, overlooking the water that epitomizes the fashionable urban trend of "warehouse living" (Storm 2008). In sharp contrast, the destroyed nuclear power plant of Chernobyl and the almost abandoned area surrounding it exemplify how still dangerous industrial environments can be considered exciting, attracting adventurous tourists in larger numbers. Finally, Karosta, a former marine base on the Latvian coast, today is slowly falling into decay after the Soviet navy left in the early 1990s (Nilsson, forthcoming). The physical remains from both the tsarist and the Soviet period constitute unused mnemonic devices for large numbers of people who are no longer present. Instead, different actors in the neighbouring city of Liepaja are struggling to take control over the future use of the area.

To a varying degree, all of these sites exemplify difficult heritage processes. Some have been articulated and recognised as heritage sites, some are undergoing gentrification processes, while others are still largely contested or even threatened with both physical and symbolic erasure. A process of reuse does not automatically imply a heritage process, but a conscious interaction with the physical remains nevertheless constitutes an arena for activating potential cultural tools. However, sites characterised by abandonment and decay could be as important as reused ones, as facilitators in a cultural heritage process, since it is the meaning attached to them that gives them value. If used in some way, physically or symbolically, they can become cultural tools for memory and heritage work.

Are the pit in Malmberget and the Ignalina plant to be looked upon as examples of the same kinds of phenomena as the above-mentioned sites? How could our views be steered in new directions by changing the context from a unique local situation to a widespread trend to reconsider industrial and military remnants as significant cultural tools for heritage processes? By framing these sites as elements in heritage processes, the pit and the plant immediately emerge as triggers of meaning and value and not only as dangerous material problems.

Nevertheless, although mining and nuclear power in different respects have been included in canonised historical chronicles and museum exhibitions, and although there are tourist mines and both operating and closed nuclear power plants that welcome visitors, it is apparently still not easy to consider these sites as connected to heritage processes. The pit in Malmberget and the Ignalina plant are spectacular but they are also most difficult to manage. What is it actually possible to do, in order to make them available as tools in people's "acts of remembering" (Smith 2006)? A first step, as I have argued in this article, would be to articulate their value, turning them from anti-landscapes into landscapes again, supporting life in a broad sense.

5. Conclusion

Two landscapes of industrial waste – what could they tell us? As the above glimpses of their rich history show, they can tell, among a number of other things, about risk, about characteristics of company towns, about generational memory shifts, about pride and fear, about dramatic changes in dependence, and economic and political preconditions that both affect future prospects and influence ways of thinking about the past.

Risks in Malmberget are visible and tangible – a huge hole, a pit, to fall into – but at the same time hidden behind vegetation and several generations of fences. In Ignalina, the risks are mostly invisible since radioactivity cannot be experienced through senses. Concerning the physical structures, on the other hand, the Malmberget pit is almost an anti-structure, something that has disappeared physically, a place of absence. In Ignalina the built environment resembles other industrial sites, and its chimneys and silhouette are a familiar part of the local skyscape. In other words, in the cases of risk visibility and physical structures, the pit and the plant have disparate characteristics.

However, similarities are easy to find. Both sites do exist solely because of the mine and the plant respectively. The local societies are intimately connected with and have long depended upon their industrial activities. In spite of the risk, one do not find expressions of fear in the local communities, probably because of their very close relationship with the mine and plant. You do not bite the hand that feeds you. Instead, there is a lot of pride, along with loyalty. However, the fear exists in other places, most prominently in the case of nuclear power. If, for example, the Danes were afraid of Swedish nuclear power, the Swedes were afraid of Soviet nuclear power, as was the EU (Storm 2010).

Furthermore, the pit and the plant carry strong narratives and stand in the centre of local attention. Yet at the same time, they are not recognized as cultural tools or mnemonic devices that are relevant in heritage processes. This is the case both for official authorities at different geographical levels, and for the local communities themselves, where I did not find any articulation "from below" concerning these issues. One explanation for this lack of articulation might be – along with the danger and risk – that both places represent not only something past but also something ongoing and part of contemporary life.

Heritage processes do not always stimulate identity and orientation, and they do not always defuse and reconcile difficult or contested memories. Heritage processes are also about power, about collective memories being negotiated, about struggles over which narratives to be put forward and which physical structures to preserve. In a sense, heritage can be regarded as a double-edged sword; to include something in a heritage process is to say both that it is valuable and that has become part of history and is therefore dead in some way. Who wants to live in a museum? The Malmberget pit and the Ignalina plant are characterized by their orientation towards the future. They both represent visions of progress and modernity, and within their respective national and ideological contexts they each once seemed "ideal" welfare communities. Today, the future is most uncertain regarding waste management as well as the prospects for the communities themselves. This is yet another circumstance that contributes to the difficulties in articulating value in relation to heritage in such places.

If the pit is ever filled, no one will be able to build on that land in a foreseeable future, and the radioactive waste management of Ignalina will be its heritage for 100,000 years. These landscapes are in a sense wounded – anti-landscapes – but by means of heritage processes they stand a chance to heal. Scars will in all probability remain, but scars carry meanings; they tell the stories that are the fabric of heritage. And so, landscapes of waste can reemerge as new landscapes, marked or even embellished by their scars.

Bibliography

Cinis, Andis, Marija Drėmaitė and Mart Kalm. 2008. "Perfect representations of Soviet planned space: Mono-industrial towns in the Soviet Baltic Republics in the 1950s-1980s" in *Scandinavian Journal of History*.

Dawson, Jane I. 1996. *Eco-nationalism: Anti-nuclear activism and national identity in Russia, Lithuania, and Ukraine*. Durham: Duke University Press.

Feldmann, Beate. 2008. "Gropen och gettot – om kulturarv, identitet och generation", in *Malmberget: Strukturella förändringar och kulturarvsprocesser – en fallstudie*, eds. Birgitta Svensson and Ola Wetterberg. Stockholm: Riksantikvarieämbetet.

Gibson, Lisanne and John Pendlebury. 2009. "Introduction: Valuing historic environments" in *Valuing historic environments*, eds. Gibson, Lisanne and John Pendlebury. Farnham: Ashgate.

Gilloch, Graeme. 1997. *Myth and metropolis: Walter Benjamin and the city*. Oxford: Polity.

Graham Brian and Peter Howard. 2008. "Introduction: Heritage and identity" in *The Ashgate research companion to heritage and identity*, eds. Graham, Brian J. and Peter Howard. Aldershot: Ashgate.

Isacson, Maths. 2004. "Kom, se dig om och förfäras! Om industriarv, museer och det berättigade i att slå mynt av vår mörka historia, 'horror tourism'" in *Kulturmiljövård* no 2.

Jönsson, Lars-Eric and Birgitta Svensson eds. 2005. *I industrisamhällets slagskugga: Om problematiska kulturarv*. Stockholm: Carlsson.

Logan, William and Keir Reeves eds. 2009. *Places of pain and shame: Dealing with "difficult heritage."* London: Routledge.

Macdonald, Sharon. 2009. *Difficult heritage: Negotiating the Nazi past in Nuremberg and beyond*. London: Routledge.

Nilsson, Louise. Forthcoming. "A past left out in the cold: An essay about the reuse of a naval base in Eastern Europe" in *Industrial Heritage around the Baltic Sea*, eds. Andis Cinis, Maths Isacson, Anders Lundgren, Marie Nisser.

Ninjalicious, 2005. *Access all areas: A users's guide to the art of urban exploration*, Toronto: Infiltration.

Olshammar, Gabriella. 2008. "Ruinlandskapet som besvär eller historia", in *Malmberget: Strukturella förändringar och kulturarvsprocesser – en fallstudie*, eds. Birgitta Svensson and Ola Wetterberg. Stockholm: Riksantikvarieämbetet.

Olsson, Krister and Anna Storm. 2008. "Det förflutna i framtidens Malmberget", in *Malmberget: Strukturella förändringar och kulturarvsprocesser – en fallstudie*, eds. Birgitta Svensson and Ola Wetterberg. Stockholm: Riksantikvarieämbetet.

Šliavaitė, Kristina. 2005. *From pioneers to target group: Social change, ethnicity and memory in a Lithuanian nuclear power plant community*. Lund: Lund University.

Smith, Laurajane. 2006. *Uses of heritage*. New York: Routledge.

Storm, Anna. 2008. *Hope and rust: Reinterpreting the industrial place in the late 20th century*, dissertation. Stockholm: KTH.

Storm, Anna, 2010. "Nuclear power plants as memory sites" *Baltic Worlds* no 4.

11

Reinventing New York's High Line

Nikolet Jensen

New York struggled for decades over what to do with an abandoned elevated railroad line. After it was slated for destruction, community activists convinced local residents, developers, and the city government to transform this rusting and dangerous anti-landscape into a green artery that afforded spectacular views of the city and it has rejuvenated the area.

Keywords: Foucault, Michel; heterotopia; High Line; historic preservation; Jacobs, Jane; Moses, Robert; New York City; railroad

1. Introduction

Anti-landscapes need not be large, obvious, or located far from populated areas. They can be abandoned industrial sites in the middle of a city, slowly crumbling to the point that they are unsafe. New York City's High Line exemplifies such a site and how it may be recovered. Today it is an elevated park on the West Side of Manhattan, close to the Hudson River. The green corridor runs for one and a half miles from the corner of West 14 and Gansevoort Street, through the Meatpacking District, Chelsea, and the Clinton/Hell's Kitchen neighborhoods, to the rail yards on West 34th Street.

It was originally an elevated industrial railroad built in response to the train accidents that often occurred along Tenth and Eleventh Avenues, where street level trains carried freight in and out of the city's largest industrial area. The commercial role of the street level trains was vital for the district; however, they also posed a hazard for residents. Due to the frequent accidents, Tenth and Eleventh Avenues were each dubbed "Death

Avenue," and in 1849 men on horses, the "West Side Cowboys," were hired to ride in front of the trains waving red flags to signal pedestrians about the danger of the moving trains. It was not the trains' speed that took victims. The streets with trolley lines were far more dangerous. According to a West Side resident interviewed by *The New York Times* in 1909, "The railroad cars run through Eleventh Avenue at snail's pace, not more than three miles an hour. They are solid trains, long enough to be seen by all…boys swarm up the cars in droves, especially if it is a coal train, to steal a ride or steal coal. In their haste to grab as much coal as possible they slip and fall and are caught. Then more victims to Death Avenue are counted." (*New York Times*, June 15, 1909)

Many immigrants came from small towns or rural areas, and they were accustomed to letting their children play on the streets. (*New York Times*, Nov 1, 1908) Accidents kept "Death Avenue" in the public debate until 1929 when the City and State of New York and the private New York Central Railroad together allocated 150 million dollars (over 2 billion today) to build the High Line, which opened in 1934. It eliminated 40 dangerous street-level crossings, and the trains increased their speed and became more efficient. The viaduct ran from West 34th Street to St. John's Park Terminal downtown at Spring Street. Over 180 buildings were torn down to make way for the elevated tracks, which ran through the center of blocks and required "some unusual engineering feats" (*New York Times*, Nov 29, 1931) such as passing through the Bell Telephone laboratory building between its second and fifth floors. The trains carried their freight directly inside some warehouses. Today, the High Line passes through the Chelsea Market building and the Standard Hotel. The line was profitable until the 1950s when interstate trucking became dominant, and it closed in 1980 when its last cargo was three carloads of frozen turkeys. The iron structure remained abandoned for a quarter century.

2. Early Preservation Efforts

A community group that called itself the Friends of the High Line saved the derelict railroad from destruction. As early as 1981, a young architect and Columbia University professor, Steven Holl, proposed turning the abandoned elevated tracks into a "Bridge of Houses." He thought that "re-use rather than demolition of the existing bridge, would be a permanent contribution to the character of the city." (Holl, 1981) The project included both mixed use houses built astride the tracks and "an elevated public promenade, and was inspired by West Chelsea's changing character from a warehouse district to a residential area. The proposal was published in *Pamphlet Architecture*, a venue for a young generation of architects. The character of the magazine reflected the audacity of his idea.

The city was not financially ready to embark on grand architectural projects, however, for it had lost 45 of the 125 Fortune 500 corporations once headquartered there. In the late 70s it also lost 600,000 industrial jobs, and more vanished in the 1980s. After narrowly avoiding bankruptcy in 1975, New York's expenditures were strictly controlled by the city administration struggling to balance its budget. (Weil, 2004, 263-267) Despite this somber financial climate, Peter Obletz struggled to convince the city of the structure's worth in a long court battle that delayed the High Line's pending demolition. Obletz, a dance-company manager and a railroad enthusiast, lived near the High Line and sometimes walked on the abandoned structure. Almost no one else then viewed the city from this unusual vantage point. Most pedestrians saw only an old useless structure hovering over the streets. The viaduct was especially frustrating for local businesses and property owners, who complained that they were "living every day with the problem of this thing just standing there, occasionally shedding concrete or steel. It's dangerous, and we feel the only thing to do with it is demolish it." (Lobbia, 2001)

In 1983, Conrail, the track owner, decided to abandon the line entirely. Obletz offered to buy it, relying on a clause in the Northeast Rail Service Act (intended to protect freight service), which stipulated that if both a track owner and the local municipality abandoned track, then within 90 days private ventures could present plans for its operation as a railroad. Obletz not only proposed running the High Line as a scrap service, carrying freight out of the city, but also proposed the High Line as a public place "when the trains weren't running" to "be devoted to some recreational uses for the community – a linear park or jogging paths, something like that." (Cook, 2009) Obletz focused on the freight part of the plan to save the viaduct from the wreckage ball, and in October 1984 acquired the rights to operate it for 10 dollars. The obsolete line was an obstacle to West Side real estate development, however, and local property owners kept Obletz in court for the next five years, during which he lost his financial supporters and the companies that had pledged to use his freight service. The city revoked his railroad rights and the High Line again faced demolition. What probably saved it was the cost of destroying almost a mile and a half of concrete and iron. The owner of the High Line (Conrail) had to furnish 7 million dollars for its destruction, and local property owners were to cover costs exceeding that amount. This may not seem a large sum in terms of New York City real estate development, however it was not an accurate one. The estimated cost of the destruction was between 20 and 30 million dollars and many feared the cost could only increase once the demolition started. (David and Hammond, 2011, 11) Before demolition could begin, the property owners had to put up a money guarantee, which they did not have. Their land under the High Line was still classified industrial, with a corresponding low valuation. (After the High Line became a successful park, the land was rezoned as residential and its value skyrocketed.)

The High Line remained untouched during a decade of legal gridlock. Finally, in 1999 the city, together with Chelsea Property Owners filed a 150 million dollar lawsuit against Conrail. Jerome M. Hauer, director of the Mayor Giuliani's Office of Emergency Management, declared Conrail's refusal to demolish the railroad 'irresponsible and negligent'', adding that it was "only a matter of time before someone is killed by falling debris. They (Conrail) can either pay for the demolition or pay the widow of the person killed.'' (Kilgannon, 1999) The administration saw the High Line as "a grim memorial to New York City's faded industrial might," (Lueck, 1999) rather than a public space with any civic value. It was an urban anti-landscape that grew more dangerous with each passing year. But then a chain of events in 1999 and after completely changed the perception of and attitude towards the iron viaduct.

3. Friends of the High Line

In 1999, Greenwich Village resident Robert Hammond became involved. He had seen the iron arches of the viaduct from the street level and it was one of the things he liked about the Village yet he "didn't even know what it was." (Greenberg, 2003) Hammond assumed that there was a group working to preserve the structure and decided to join as a volunteer. He went to a few community meetings and to his surprise realized that after Peter Obletz had passed away in 1996 no one was working to keep the structure standing. Hammond met Joshua David, who also wanted to join a group working for the High Line's preservation, and they founded The Friends of the High Line. At first, according to Hammond, "it was a project to work on and an excuse to get permission to go up on the line." As soon as Hammond began to walk on the High Line he was convinced that the structure was worth fighting for: "Once I got up on top I realized there's a whole world up there. It's like a meadow right in the middle of

All photographs by Nikolet Jensen, 2011-2012.

the city. I've always loved the juxtaposition of a wildflower meadow on a steel structure in the middle of Manhattan with a view of the Empire State Building." (Greenberg, 2003)

Meanwhile, the rail and shipping company CXS acquired the bankrupt Conrail and took over the High Line. The annual city taxes of 400,000 dollars for the obsolete viaduct prompted CXS to look for the fastest and cheapest way to "get out of the High Line business," (Cited in Hammond, 2011, 11) which included taxes, maintenance costs, and liability. It did not matter to the corporation if it was converted to a park or demolished, and they negotiated with both sides. In December 2001 in the last days of Giuliani's administration, an agreement for the High Line's removal was reached. However, the Friends of the High Line had already spent two years gaining support for the structure's salvation. Their vision to convert the old railroad into an elevated park similar to the Promenade Plantée in Paris had won the attention and support of New York's succeeding mayor Michael Bloomberg.[1] He reversed the demolition decision and supported the so-called "Rails to Trails" designation, a certificate of interim trail use, which would allow the city to treat the railroad as a public space. The tracks could be used for other purposes as long as the structure was kept intact and it could be converted back to a railroad, at least in theory. The High Line was not the only applicant for such designation in the United States. The first abandoned railroad was approved for trail transformation in 1986. By 2000, over 11,000 miles of obsolete tracks had been converted into public trails and green corridors for walking, bicycling, and cross-country skiing. (Freierman. 2000) However, the High Line was different given its urban setting and elevated structure. Most designated trails were already surrounded by nature and could easily become parks, but the High Line was an iron structure surrounded by concrete and warehouses. The only green nature in proximity was the wild flora growing between the rusty rails. It was a risky and costly decision for the city to invest in a promenade, which was estimated to cost over 100 million dollars. The industrial tracks, erected amidst the city's worst financial depression, and closed after New York's near bankruptcy, were now up against the financial recession in

1. It helped that Robert Hammond's college roommate was the City Council speaker during Bloomberg's administration, or the second highest elected official in New York City. Greenberg, 2003.

the aftermath of September 11. However, the city also needed hope and regeneration after the destruction and desperation of 9/11.

In 2001, as the positive attitude towards the High Line project continued to grow, Hillary Clinton became a New York State Senator. She was already experienced with designating old tracks as recreational paths. In 1998, the then First Lady, together with an organization called Rails-to-Trails, announced the opening of 16 National Millennium Trails to "honor the past and imagine the future" and "to spark the creation and enhancement of more than 2,000 trails as part of America's legacy for the future."[2] The High Line fitted this narrative, so Mrs. Clinton's support was not surprising. In 2005 she walked on the High Line to celebrate the passage of a federal transportation bill allocating $18 million for its conversion into an elevated park. (Amateau, 2005) The High Line clearly enjoyed political, public, and media support. However, it is interesting to trace the initial efforts of Friends of the High Line to see how the viaduct gained so much attention and praise.

4. The High Line Project

In the early 1980s Peter Obletz seemed to be almost alone in his fight for the tracks' survival. His personal fascination with railroads earned him the media reputation of "train buff", which diluted his true intentions of using the High Line as a public space, rather than a cargo line. However, 20 years later, the trains on the High Line were just a part of its historic narrative. Its preservation evolved around other ideas. By the early 2000s, the neighborhoods bordering the High Line had undergone tremendous real estate development. Chelsea and the Meatpacking district had become some of the most popular new residential areas in the city. Chelsea had also established itself as an art district. Its vast industrial lofts now accommodated modern art galleries. The Meatpacking district had also experienced a revival with posh restaurants, fashion boutiques, and new residents moving into the remodeled warehouses. The High Line project seemed to continue the process of reinvention already shaping the area.

2. First Lady Hillary Rodham Clinton, U.S. Transportation Secretary Slater Announce 16 National Millennium Trails, White House Mullennium Council, June 26, 2000, http://clinton2.nara.gov/Initiatives/Millennium/trails_doc.html.

However, as recently as the late 1990s many New Yorkers were not aware of the derelict railroad's existence. Even if they had seen it from the street level, they did not think much about it. The qualities of the viaduct were only revealed when one climbed atop of the structure. The abandoned railroad was closed to the public and only a small number of people had walked on it. That is why the Friends of the High Line decided to bring the railroad to the public. Hammond and David knew the power of publicity in the city. In March 2000 the group convinced Joel Sternfeld, a well-known American photographer, to document the High Line in photographs. Hammond knew that in order for the preservation project to succeed they "needed people to see what was up there." (Monchaux, 2005) In 2001, the group arranged a benefit in the Mary Boone modern art gallery in Chelsea. The High Line activists distributed postcard invitations with a photograph of the High Line wilderness. One of these postcards got in the mail of actor Kevin Bacon. He later said that when he saw the picture he could not believe this green space truly existed on Manhattan. (Carlson, 2009) His appearance at the benefit helped the High Line gain media publicity and exposed more New Yorkers to the rough beauty of the abandoned tracks.

Attracting celebrities such as Bacon, fashion designer Diane Von Furstenberg, actor Edward Norton, and TV and business mogul Martha Stewart, helped the Friends of the High Line receive media attention. However, it is interesting to note that these people did not join the battle randomly – Edward Norton's father was the first chairman of the Rails-to-Trails Conservancy, Von Furstenberg was already in a community fight to preserve the neighboring Ganservoort Market, Kevin Bacon's father was a city planner and author of "Design of Cities," and Martha Stewart's offices overlooked the High Line. They brought the media spotlight as well as financial support through fund-raisers. If one looks at the High Line press archive, the amount of publicity grows with each year. In 2001, the project was described in two newspaper articles. In 2002, there were eight, in 2003, twenty, and in 2004 there were even more. The avalanche effect was clear. However, the question remains why the interest in the preservation of the old viaduct spiked so dramatically when it lacked most of the characteristics of something worth preserving. It was not educational, nor aesthetically beautiful, and its destruction would displace neither people nor businesses. The opposite was the case; destruction would have created more space for real estate development.

However, despite lacking many of the traditional reasons for preservation, the industrial past of the High Line and the images of its abandoned wilderness, coupled with the city vistas it revealed, spurred people's imagination. In 1972, urban planner Kevin Lynch wrote that "the contrast of old and new, the accumulated concentration of the most significant elements of the various periods gone by, even, if they are only fragmentary reminders of them, will in time produce a landscape whose depth no one period can equal…we look for a setting that, rather than simply being a facsimile of the past, seems to open outward in time." (Lynch, 1972, 57). The High Line provided such settings and provoked the re-imagining of New York's urban landscape. In 2003, a Friends of the High Line contest for redevelopment ideas received 720 different proposals from 36 countries. The following year, 52 teams of prominent architects, engineers, and landscapists competed with plans for the 1.5 mile project. Among the finalists was the team of architect Steven Holl, who had proposed the Bridge of Houses plan in the early 1980s. Now, his vision of the tracks was "as a suspended green valley in the Manhattan Alps." (High Line Official Website, 2011) The residential possibilities of the High Line were omitted and the emphasis was on its recreational use. Perhaps the most renowned finalist was the team of Zaha Hadid, who focused on the High Line's cultural aspect. The tracks would accommodate artistic events on various stages and event platforms. Their re-design was inspired by a Möbius strip with curves and curled edges, which made the structure seem fluid.

The design winner was Field Operations (Diller, Scofidio & Renfro), whose team specializied in urban ecology and emphasized not only recreation but also "ecosystems, urban cultures and sustainable economies." The project strived to preserve "the romance of the ruin" with the naturally developed ecosystems that were already in place on the High Line. It placed importance on "how to maintain the magic of the High Line as a found landscape in the city." (Pogrebin, 2005) The winning design for the new promenade found the direction that the High Line project was striving for – a balance between nature and urban environment. It preserved the initial appeal of the abandoned railroad as a spontaneous oasis of nature and incorporated it into the city landscape revealing ideas about new urban development in New York, where nature and urban environment intertwine. In this sense, the High Line's relationship to the city is very different than that of Central Park. Central Park is an oasis of green and the emphasis is on nature. The High Line's recreational concept does not isolate one from the city. Quite the opposite, it is the city that is revealed from the elevated structure. The winning design equally combined the concrete and green elements. This way one can enjoy and re-imagine the nature that grew on top of the abandoned track, but it is also a reminder that the High Line is part of the urban environment. From the elevated structure the focal point is not only nature on the High Line but also the manmade landscape, the peaks and canyons of Manhattan. The promenade can be seen as a celebration of the city.

According to urban scholar Roger Keil, "urbanization is not merely a linear distancing of human life from nature, but rather a process by which new and more complex relationships of society and nature are created." (Kinkela, 2009, 905-928) The High Line as a project seemed to strive for a dialogue between society and different expressions of nature. The elevated railroad was part of the urban landscape, but at the same time it was isolated. Its preservation and re-design was also its reintroduction and incorporation to the city environment. The structure's vantage point offered a new relationship between the metropolis and its habitants. It also juxtaposed green nature on top with the surrounding "manmade nature."

The High Line design team included Danish artist Olafur Eliasson, who in 2008 created The New York City Waterfalls. His waterfall installations in Manhattan, Brooklyn, and Governor's Island posed similar questions about human and manmade nature. For example, the waterfall under the Brooklyn Bridge was supported by metal scaffolding, but even the artist himself did not predict how well its mechanic structure would blend in as a part of the city. In this sense an urban environment is not only planned but also accidental. The blending of nature and industrial elements begins to look strikingly natural. A new trend could be seen to emerge in which manmade environment and nature are not separated but brought together.

The High Line also challenged the way people move within Manhattan. The New Yorker is accustomed to walk through the city grid crossing streets, or avenues, in linear motion with 90 degree turns, except for Broadway and on the southern tip of the island. The High Line offered an uninterrupted walking path, which did not follow the strictly arranged grid on the street level. At the same time, the High Line walk was broken into segments, and although many refer to the structure as a linear park, its path's curves are not smooth, but choppy to create a new

way of moving. As a High Line visitor noted, part of what made the walk so interesting was the different designs of its segments.[3]

The High Line cuts directly over 10th Avenue, where people can sit in theater-like settings, looking through a clear screen as if they were in the middle of the busy avenue below. The promenade's design also includes a narrow, tree-lined path, squeezed between old brick buildings, which then opens up into a broad area revealing the brand new high rises erected on each side of the promenade. The combination of the various segments, directs the pedestrians' movement and creates focal points, altering between layers of old and new architecture, and giving a continuous sense of discovery. This makes people slow their pace as they walk and gives them time to reflect and enjoy what they see on their way. At the same time the elevated park kept its old rail tracks as part of its character. This way the pedestrians are reminded, in a subtle manner, of the busy industrial past when trains were rolling above the Manhattan's West Side.

The High Line could be seen as what the French philosopher Michel Foucault described as heterotopia or a space that can have many, sometimes contradictory meanings at the same time. In his text, "Of Other Spaces," (1967) he identified several different types of heterotopias including "heterotopias of indefinitely accumulating time, for example museums and libraries" where in a single space one can observe the build up from different times. The High Line chronicles New York's industrial past which now is woven into the structure's re-development. The railroad tracks incorporated in the redesigned viaduct, keep the connection with its original purpose in early twentieth century industrial New York. Another heterotopian aspect the High Line embodied was the "juxtaposing in a single real place several spaces, several sites that are in themselves incompatible." One of Foucault's primary examples of juxtaposing of different places was the garden. In the Orient the garden had a rectangular shape representing the four parts of the world and therefore, encompassing its totality. (Foucault, 1984) As blogger Gala Lutteroth notes, the High Line is a garden and in this sense could be seen as Foucault's "universalizing heterotopia that allows the whole world to fit in to a reduced space." (Lutteroth, 2011) Going beyond the theoretical, the promenade often feels like a juxtaposition of various parts

3. Interview with a High Line visitor on Nov.14, 2011.

of the world in one narrow stretch. With over 2 million visitors since its opening, one could hear almost any language spoken in the elevated park. As National Geographic noted, "you are as likely to hear visitors speaking German or Japanese as English." (Goldberger, 2011)

The High Line's new design also juxtaposed the worlds that inhabited the High Line at different times. After the tracks had closed for business, the abandoned structure took on a life of its own. Hidden from people's eyes, hovering over one of the busiest cities in the world, the rusty tracks seemed to be enveloped in silence and wild flora that steadily grew among the tracks. The viaduct turned into a forgotten world, a quiet life that countered the urban reality underneath. The under-the-radar nature of the derelict space provoked "spontaneous and imaginative reactions from individuals and groups." (Arroyo, 2010) The High Line's accidental transformation from a railroad into a mystified world showed that the urban landscape is still hard to control and that it can take a direction of its own. The tracks' evolution suggested that "other things can still spontaneously develop in the city: subcultures, neighborhoods, ideas." (Stalter, 2006, 869-890.)

Such heterotopias juxtaposing time periods and function provoked people's imagination and made the physically robust structure of iron and concrete into a very flexible concept in people's minds. This flexibility made the project accessible to a broad public and gained wide support for its preservation and redevelopment. In a way, the structure became a Rorschach test – people saw it differently – a park, a railroad, a green corridor, a new vantage point, a preservation project, a timeline of New York's development, a business opportunity, a historical heritage, a forgotten world, a garden. The flora growing among the rusty tracks attracted the interest of landscapists, biologists and horticulturists. Some of the species growing there could not be found anywhere else in the city. At the same time, the High Line was attractive to architects challenged by its industrial aesthetics. Others saw it as a romantic artifact. Whatever the interpretation, the old tracks proved to be a well of inspiration.

5. The High Line Inspiration

After its "discovery" and redevelopment, the High Line inspired new ideas beyond its borders. One New York woman used her friend's balcony overlooking the High Line to sing to pedestrians in an act that she called Renegade Cabaret. (*New York Times*, 6.17.2009) A dance company staged a performance on nearby rooftops designed to be seen from the High Line. New York's Whitney Museum pledged to build a branch near the elevated park, to be completed in 2015. The district is close to the Chelsea art galleries, and the new museum will help establish a new cultural aura for the area. At the moment, most New York museums are concentrated on the Upper East Side of the city. Promenade visitors can not only witness spontaneous and organized art, but also participate in the shaping of the High Line's identity, for it has become a runway for New Yorkers' fashion as well as a relaxed community garden. A roller skating rink has opened underneath the High Line, where an old parking lot used to be. An outdoor bar has opened below the structure on 30th Street. High Line ventures range from picnic baskets taken along on the promenade to extravagant buildings constructed in the newly emerged neighborhood. The new cultural energy has spilled outside of the tracks and helped the surrounding neighborhood flourish. What had been long been perceived as a dangerous and deserted anti-landscape became the kernel of a neighborhood with a new identity, known as the High Line district.

Since its opening, the park in the sky ironically has benefitted its former adversaries – the local property owners and the city. Land values and business opportunities in the area have risen dramatically. Like the early development of the railroad, along which cities sprang up, the High Line has attracted new ventures and construction around itself. A deli owner on 27th Street and 10th Avenue found that the promenade was very good for business, attracting so many more customers that he is soon to open a second store, a block away, on 28th street, which he will name "The High Line Darling."[4] Overall, the revitalized district created 12,000 new jobs. The City of New York invested a total of 170 million dollars in the project, which has stimulated direct investments worth 5 billion dollars for as many as 50 nearby major residential projects (High Line Official Website, 2011).

Old buildings have been renovated and new buildings emerged on the tiny plots available. The new HL23 building, whose aerodynamic design overlooks the High Line, is part of the new generation of condominiums mushrooming along the promenade. The HL23 architects considered not only the promenade's vistas but also its "machine aesthetic." The building's steel surface panels, resembling car hood scoops, were manufactured on presses used to make body parts for Mercedes trucks. (Ouroussoff, 2011) Its design brought futuristic yet nostalgic urban visuality to the High Line. Incorporating remnants of New York's industrial past brought

4. Interview with a store owner on 27th Street and 10th Avenue, Nov. 14, 2011.

back memories of the early 1930s when the High Line was built and the Chrysler Building's machine-inspired aesthetics shot into Manhattan's skyline. From this point of view, the High Line has become a timeline and a link to the vanishing landscape of the industrial era.

The new zoning laws around the High Line made it possible for buildings to be erected right next to the promenade. While the old buildings along the railroad had no windows or have sealed the ones overlooking the tracks, the newly built structures compete with bigger windows and better views of the viaduct. This new interaction between the High Line and its surroundings has recreated another forgotten urban visuality – the elevated train tracks of the "El." Before it was dismantled in the 50s, the El carried passengers over Manhattan's Third Avenue, giving them the chance to see the city from above and peek into the windows of the buildings near by. The "El" presented "broad vistas that connected neighborhoods as well as intimate close ups that segmented vision into brief glimpses." (Stalter, 2006, 873) The same can now be said of the High Line. However, while the elevated path of the El revealed segments of the working class life, the High Line voyeuristic appeal is in state-of-the art, luxury apartments. New York commuters once glimpsed the poor through the windows of the El, while pedestrians on the High Line can peek into the life of the rich in the buildings surrounding the promenade. In this sense, the High Line brought back to Manhattan the forgotten experience and vantage point of the El.

The British design and architecture critic Hugh Pearman wrote that walking on the High Line "was a joy" because of the "utterly different perspective on Manhattan that it affords." According to him, "originally prosaic," the High Line "has acquired touches of the sublime." (Pearman, 2009) One can indeed feel the sublime experience when walking atop of the structure. The impressive skyscrapers on display have long been regarded as an expression of "geometrical sublime." (Nye, 1994, 77) However, most people only see them from the street level or from the top of a skyscraper observatory. Some lucky New Yorkers also have the chance to look at their city from office or apartment windows. However, these present picturesque but static frames of the city. The sublimity of the High Line views is not only in the vantage point but also in its ability to change as one moves on the promenade. The elevation from the street level allows broader vistas of the city giving the experience more depth. At the same time, one remains connected to the living, breathing city underneath. An observatory experience, from the deck of the Empire State Building for example, astonishes the visitors, however it also detaches them from the reality of the city – the streets look like small paths, the people look like ants and the cars like toys. The High Line experience encompasses both the human scale of the city and its architectural grandeur.

In this sense, the elevated park re-invents the sublimity of New York's skyline, adding new dimensions to the experience.

6. The High Line Between Moses' and Jacobs' Legacies

In 2010, The High Line's founding activists were presented with the Jane Jacobs medals for new ideas and activism given each year to recipients whose work "creates new ways of seeing and understanding New York City, challenges traditional assumptions, and creatively uses the urban environment to make New York City a place of hope and expectation." (Rockefeller Foundation, 2010) The High Line had certainly borrowed ideas from Jacobs and shown how a declining neighborhood can regenerate itself. However, there were also some curious ironies about the award. The High Line was part of one of Robert Moses' grand ideas – the West Side Improvement project. From this point of view the High Line combined both the legacy of Moses and Jacobs, whose urban visions were in conflict for many years. One could argue whether the outcome of the project owed more to the influence of Moses or Jacobs. On one side, Moses' decisions were driven by practicality, and perhaps he would have destroyed the elevated tracks to make more space for development. He might have thought of turning the railroad into an automobile ramp – a parkway carrying traffic above Manhattan's West Side. However he was also New York City's Park Commissioner, and the High Line might have caught his attention as a recreational project.

On the other hand, Jane Jacobs, who often opposed Moses' ideas, believed that neighborhoods have the means to regenerate from within. In this sense the High Line has shown its ability to re-kindle people's imaginations and drive a wave of neighborhood improvements. However, one could question whether Jacobs would have fought to keep the tracks in the city. According to her, the areas surrounding railroad tracks, or other places of "massive single uses" (Jacobs, 1992, 259) such as large parks were seldom areas that attracted people. The High Line also fits Jacob's description of "border vacuum" as a massive structure that creates a physical and mental border in a neighborhood. Jacobs also represented the middle class in the city. In the case of the High Line one could argue that the gentrification of the area affects its diversity. The rise of property values and rents as well as the proliferation of luxury condominiums drives the middle class away. Jacobs' activist record reveals that she fought to prevent the construction of luxury apartment buildings in the West Village in favor of preserving its character and diversity. Moses and Jacobs might have supported or opposed the High Line for various reasons. Both introduced innovative ideas about the city; however, neither of them could have predicted the full ramifications of their efforts. The urban environment is flexible and hard to fully control. In *Triumph of the City*, Harvard economics professor Edward Glaeser argues "cities can't force change with new buildings... But if change is happening, the right kind of new building can help that process." In this sense, the High Line could be seen as the agent of change taking shape in the city.

7. The Right Time for the High Line

Preservation projects do not work in isolation but are highly dependent on society for support. The ideals of an era often make the difference between demolition and conservation. In the case of the High Line, the elevated viaduct did not at first stir appreciation, and its qualities and industrial aesthetics were not understood in the early 80s when Peter Obletz was fighting to preserve it. Indeed, The High Line is a very new type of preservation project that combines urban preservation, historic preservation, and environmental preservation. Compared to other preservation projects, it uniquely codifies both New York's past and future. The conservationist at the beginning of the century strived for historical and educational value. In the City Hall preservation for example, the goal was to safeguard a building that reminded urban dwellers of New York's heroic past. Its educational purpose was history and patriotism. In the mid century, Jane Jacobs's preservation battles were aligned with counter narratives that challenged the established ideas of top-down urban planning. The preservation of SoHo had the unexpected outcome of creating one of New York's most vital and popular districts that combines the beauty of the cast iron architecture with the modernity of the boutiques and restaurants dotting the neighborhood. Her preservation battle prevented the displacement of people and businesses as well as preserving the aesthetic qualities of New York's architectural past. Yet similar efforts failed to preserve both The Singer Building and Penn Station, and many other structures also vanished from the cityscape.

Compared with these iconic structures, the High Line's survival seems almost miraculous, but it was due to a combination of legal complexities and high costs that delayed action. In addition, the focus was not on preserving the structure's original purpose. The railroad tracks were an important part of the park's novelty, but the focus was not on creating a museum of New York's industrial past, but rather on reinventing the structure to provide visions of the city. At the beginning of the twentieth century, Manhattan propelled itself by destroying the old and building the new. The city kept its cutting edge by razing the past and quickly updating its landscape. The High Line reveals a contrasting approach. Instead of being demolished, the city environment is reinvented. The project uses an old, mechanically sound structure to create a new recreational area in a part of the city that has lacked open space and broad vistas.

The park benefits not only the neighborhood but also the city. Open spaces in New York are scarce compared to other metropolises. London has 4,795 people per square kilometer, and open, green spaces comprise 38% of the cityscape. New York's density is 24,000 people per square kilometer, and there is less than half as much green, open space (14%). (Burdett and Sudjic, 2006, 76, 140) This makes the High Line a very important project in terms of quality of life, recreation and sustainability.

The High Line expresses a new era of environmental consciousness and concern about the future urban development. It promotes sustainability, an idea that has dominated the public agenda in the new century. The High Line commemorates New York's industrial past yet it also leads into a greener future with more open spaces. The project's intertwining of nature and built environment helps New Yorkers imagine their future. It exemplifies how an anti-landscape can be acknowledged, recovered, and transformed.

Bibliography

Amateau, Albert. 2005. "Hillary hikes on the High Line and hooks up $18 million for it," *The Villager*, August 10-16.

Arroyo, John C. 2010. "Culture in Conrete: Art and the Re-Imagination of the Los Angeles River as Civic Space." MA thesis, MIT.

Burdett, Ricky, and Deyan Sudjic. 2007. *The Endless City*. London: Phaidon Press.

Carlson, Jen. 2009. "Kevin Bacon Talks the High Line," *The Gothamist,* May 26.

Cook, Kathy. 2009. "The High Line: Honor Peter Obletz!," *Cook's Log Blog*, June 27, http://cookslogblog.blogspot.com/2009/06/high-line-4-honor-peter-obletz.html

David, Joshua and Robert Hammond. 2011. *High Line: The Inside Story of New York City's Park in the Sky*. New York: Farrar, Straus and Giroux.

Foucault, Michel. 1967. "Of Other Spaces (1967), Heterotopias," *Architecture Mouvement Continuité*, October 1984, http://foucault.info/documents/heteroTopia/foucault.heteroTopia.en.html.

Freierman, Shelly.2000. "From Rails to Trails, With Miles and Miles to Explore," *The New York Times*, June 29.

Goldberger, Paul. 2011. "New York's High Line: Miracle Above Manhattan," *National Geographic*, April, http://ngm.nationalgeographic.com/2011/04/ny-high-line/goldberger-text/1.

Greenberg, Mike.2003, " S.A. native works to save N.Y.'s High Line," *San Antonio Express-News*, October 26.

High Line Official Website, 2011. Press Archive, accessed Sept.13, http://www.thehighline.org/blog/tags/crains-3.

Holl, Steven. 1981. *Bridge of Houses*, Pamphlet, Architecture 7, 1981, William Stout Books, http://www.stevenholl.com/books-detail.php?id=52, accessed Sept. 2, 2011.

Jacobs, Jane. 1992. *The Life and Death of Great Cities*. New York: Vintage Books.

Kilgannon, Corey. 1999. "Fight Over Unused Rail Line," *The New York Times*, May 16.

Kinkela, David. 2009. "The Ecological Landscapes of Jane Jacobs and Rachel Carson," *American Quarterly*, 61:4, Dec.

Lobbia, J. A. 2001. "One Track Mind, Chelsea Group Works to Save an Abandoned Rail Line," *The Village Voice*, Dec. 27. http://www.thehighline.org/press/articles/122701_voice/

Lueck, Thomas J. 1999. "Up, But Not Running on the West Side," *New York Times,* July 25.

Lutteroth, Gala. 2011. "High Line New York," *Urban Media Lab Waste*, Feb. 9, http://urbanmedialabwaste.wordpress.com/2011/02/09/high-line-new-york/.

Lynch, Kevin. 1972. *What Time is This Place?* Cambridge, MA: MIT Press.

Monchaux, Thomas De. 2005. "How Everyone Jumped Aboard a Railroad to Nowhere," *New York Times*, May 8.

New York Times, 1909. "Safety on Eleventh Avenue. Old Resident Explains Its Unmerited Reputation for Accidents," June 15.

New York Times, 1931. "West Side Improvement Project Advancing Rapidly," Nov. 29.

New York Times, 1908. "That is a Much Higher Rate than Is Shown by Europe's Great Cities, Where Safety and Comfort Rather than Speed and Capacity are Traction Rules," Nov. 1.

New York Times, 2009. "The Renegade Cabaret," June 17.

Nye, David E. 1994. *American Technological Sublime.* Cambridge, MA: MIT Press, p. 77.

Ouroussoff, Nicolai. 2011. "Nostalgia Wrapped in Steel," *New York Times*, August 25.

Pearman, Hugh. 2009. Manhattan's High Line: I think it's going to be OK," Gabion, Feb. http://www.hughpearman.com/2009/06.html.

Pogrebin, Robin. 2005. "Designers Detail an Urban Oasis 30 Feet Up," *New York Times,* April 19.

Stalter, Sunny. 2006. "Farewell to the El: Nostalgic Urban Visuality on the Third Avenue Elevated Train," *American Quarterly*, 58:3, Sept.

Weil, Francois. 2004. *A History of New York.* New York: Columbia University Press.

Afterword

12

View from the Dump: Stige Ø and the Question of Anti-Landscapes

Sarah S. Elkind

The Stige Ø area on the edge of Odense fjord exemplifies a recovered landscape. A modest community of workers and fishermen were evicted from the site to make way for a dump, which has since been closed and transformed into a recreation area. What do such sites suggest about the concerns of the papers in this volume, or about the idea of anti-landscape itself?

Keywords: Anti-landscape; land reclamation; Odense, Denmark; recreation; solid waste

1. Introduction

North of Odense, a narrow peninsula separates the city's shipping canal from the mouth of the Odense River and the rest of the estuary called the Odense Fjord. At the base of the peninsula, apartment blocks rise where huge grain silos once stood in the all-but abandoned harbor. A municipal incinerator and central heating plant looms over the cottages in a community garden and a sailing club. Further along the peninsula, a winding road enters a stand of young trees, and then the landscape opens on a hilly park, the site of circus performances, a mountain bike course, creative climbing structures, camping shelters and environmental art installations. Even further north, the land flattens out again to reveal a pasture and a cluster of small cabins which once housed a small community of fishers, hunters and shipyard workers.

Stige Ø was always a marginal place. In the early twentieth century, the peninsula attracted visitors with a dance hall, diving exhibitions, circus performances and puppet shows. For sixty years, a small shell industry,

Mortensens Skalleværk, harvested clam and mussel shells for chicken feed. In the 1960s, Odense moved its municipal dump from the eastern outskirts of town to Stige Ø to make room for suburban housing. Over the next thirty years, 10 million cubic meters of domestic and industrial waste from Odense and several nearby cities expanded the peninsula by 54 hectares. Waste disposal affected fishing in the fjord, attracted hordes of scavenging seagulls, and let loose a blizzard of wind-blown plastic debris on nearby communities. Then, local officials closed the Stige Ø dump, and moved waste disposal to a new landfill across the canal with a greater capacity and better provisions for controlling leachate and other environmental impacts. (Larsen 2010).

The reconversion of Stige Ø from landfill to landscape began in 1994. In its current incarnation, the peninsula embodies the anti-landscapes discussed elsewhere in this volume. It is a hybrid, reclaimed space: a twenty-first century "nature" park on an abandoned twentieth century landfill placed on a manmade peninsula that was itself created out of spoils from dredging the Odense canal in the late eighteenth and early nineteenth centuries. Like other anti-landscapes, Stige Ø is the legacy of modernism. The visible industrial legacies include the land itself, constructed from the capped refuse of industrial production and consumption. The invisible legacies include leachate and methane from thirty years of accumulated and decomposing waste. Tent-camping is prohibited on Stige Ø because this methane might accumulate and suffocate tent-campers in their sleep. A system of pipes collects and transports the leachate to a treatment plant to prevent contamination of the fjord or, by extension, the Baltic Sea (Opsamling af Forurenet Regnvand), 2013.

2. Anti-Landscapes and Cultural History

David E. Nye opened this volume with a definition of anti-landscapes that emphasized their material character. If landscapes provide "the infrastructure of collective existence," then an anti-landscape is a space that has ceased to support humankind (Nye 2010, 130-31). The processes that reduce a landscape to an anti-landscape are closely linked to modernism, industrialism and consumption. American environmental historians, incorporating theories of place into their investigations of the nexus of human society and nature, have taken this dynamic one step further. They argue that capitalism – at least American capitalism – is so inherently alienating that it always strips the cultural foundations of place from the landscape (Flores 1994, 15; Worster 1979). If they are correct and culture is always grounded in regional distinctiveness as well as environmental sustainability, every place will eventually become an anti-landscape no-place.

Several authors in this volume address the question of environmental sustainability, depletion of natural resources, or the creation of environments too toxic or sterile to sustain human life. But most treat anti-landscapes in Worster's and Flores's cultural terms. Anti-landscapes for these authors are not just places that cannot sustain life, but also places that have become culturally sterile or metaphorically invisible. These works invert the relationship between place and culture by considering places that *could* sustain community and culture but no longer do. In some cases, a group of people left or an industry closed, taking the culture that gave that place meaning with it. In others, soulless development impaired community formation or kept a location from acquiring cultural significance in the first place. These anti-landscapes arose less because a place changed, than because an economy or community did.

Heritage, tourism or recreational development can reclaim these cultural anti-landscapes, after a fashion. Thus, capping the Stige Ø landfill, installing camping shelters, picnic tables, play structures and paths restored the peninsula as both a recreational destination, and a potential foundation for community-building. In New York City, the High Line park project turned the abandoned and invisible rail viaduct into a community asset. This type of restoration is relatively easy: creative remediation and landscaping (pun intended) can make recreational spaces out of many sorts of uninviting places. Even more frankly dangerous spaces such as Chernobyl attracts tourists. At Ignalina or Malmberget, where physical restoration seems nearly impossible, heritage processes offer cultural, if not physical, restoration. In other places, too, the eerie beauty of sites abandoned by humans – but not by nature – in the wake of apocalyptic events attract tourists fascinated by or grappling with our shared history of industrial and natural disaster. That shared history may re-integrate these anti-landscapes into our common heritage; physical restoration, malleable culture and heritage may combine so that these spaces sustain community again.

When the loss of cultural significance or mere ugliness is the problem, the impulse to restore an anti-landscape through physical and cultural remediation is unsurprising. But by and large, reclamation has as its goal recreation rather than new industrial production. Given that industrial and agricultural production consume recreational resources as well as other ecological, cultural and economic assets, the construction of recreational places out of anti-landscapes is a valuable, if limited, public good. Communities that have lost their main resource-extractive or manufacturing industries have invested a great deal of money and hope in tourism; tourism frequently centers on creative reclamation of their industrial ruins. So, the semi-abandoned mining towns of the American West feature scenic highways, historic parks and special gold- and silver-boom tours, occasionally replete with mock gun-battles along carefully groomed frontier-town streets that look more like the back lot of a Hollywood studio than a living community. Starting in the 1930s, the anti-landscape of spent silver and gold mines in Aspen, Breckenridge and Telluride gave way to a swank ski industry that continues to capitalize on a romantic version of Colorado's mining past (Coleman 2004). As Hal Rothman has shown, however, tourism economies are precarious and problematic. Successful tourism development inflates property values until they are well out of range for locals employed in the tourism industries, and attracts national and international merchants at the expense of local entrepreneurs. Rothman and others argue that this sort of large-scale tourist development does not sustain the communities that hope it will reclaim their anti-landscapes (Rothman 1998; Long 2001).

Tourism is not the only sort of reclamation that can fail. Many ambitious restoration projects leave impoverished landscapes behind that, while no longer quite as impoverished, still do not sustain the human or animal communities that they once did. Between the 1880s and 1950s, a wealth of timber, salmon and hydroelectric energy spawned rapid industrial development in Seattle, Washington. Dams, clearcut forests and navigational improvements quickly transformed the region from harbor to hinterland. Native Americans were abruptly dispossessed; those that stayed in the Seattle area struggled as they lost access to fish, shellfish, land and their culturally important places. In the 1970s, environmentalism drove Seattle residents to preserve open space and regulate water and air pollution. They succeeded well enough to give Seattle a national reputation as a hip, green city. Nevertheless, even after fifty years the shellfish, finfish, and plant life that had sustained the Salish, Duwamish and Shilshole people have not recovered or they remain too toxic to eat (Thrush 2007, 79-105, 162-183). Hundreds of miles south of Seattle, the combination of silt and heavy metals from nineteenth-century gold mining and pollution from twentieth century industries extirpated the oyster and fin fisheries of San Francisco Bay (Booker 2006). Seattle and San Francisco Bay are still beautiful, and still the centers of lively communities. They do not look like the anti-landscapes described in this volume. But industrial production and its byproducts have long since transformed the waters at the heart of each of these regions from productive and life-sustaining resources into greatly impoverished scenic resources.

Stige Ø witnessed a similar pattern of evolution from productivity to recreation: the shell industry exhausted local shellfish beds in 1970 and closed; the local eel fishery crashed around the same time; commercial shipping abandoned the small canal and harbor in favor of new rail lines, bridges and highways that connected Odense to mainland Europe in the late twentieth century (Larsen 2010, 7). In the 1960s, the Odense government literally removed the community from Stige Ø when it ruled their houses too substandard for year-round occupation, and ordered the community of hunters, fishers and dockworkers who lived there to find safer residences elsewhere (Larsen 2010, 15-20). Reusing Stige Ø dump as a park probably represents the best use of the former landfill, but as a recreational area it will never provide the same kind of sustenance that its former fishing, shipping, and shell industries did. In this way, Stige Ø, like Seattle, San Francisco Bay, and the ski areas of the American Rockies, raises one of the central questions about anti-landscape: if an attractive place is merely pretty, but does not sustain community in other, non-aesthetic ways, it is still a landscape? To be more concrete, as a lovely and well-designed bit of open space, Stige Ø will support community-building in the Odense region. But because industrial development transformed the peninsula from a productive to a recreational resource, it may never be more than a very pretty post-industrial anti-landscape.

3. Anti-Landscapes of the Future

Together, the papers in this collection suggest a distinct life cycle for landscapes and anti-landscapes. Landscapes are made by human occupation and use, or by identification as places of aesthetic or cultural value. They are experienced as productive, beneficial spaces, sometimes as home, sometimes as places of industry. Then, they decline as landscapes due to resource exhaustion, accumulated pollution, and economic change. Sometimes this decline results from an industrial catastrophe like that which struck Chernobyl, or a natural disaster such that which followed Hurricane Katrina in 2005. Sometimes, it is a more benign process of cultural transformation through which a place simply ceases to matter. Emerging anti-landscapes may be experienced simply as places in decline, they may come to be seen as sources of disease, social decay or discomfort. Finally, they are represented as anti-landscapes of uselessness, meaninglessness, decline or danger. Then, in many of the papers here, they are reclaimed for recreation, tourism or residential development. If not, they may become unintended environmental reserves, which if the "collective existence" in Nye's definition of an anti-landscape includes animals and plants, means that even some of places most inhospitable to people may eventually become landscapes again as well.

This life cycle of decline and redemption applies only to the smaller, more compact anti-landscapes, considered without much reference to their global, environmental contexts. The cumulative impact of industrial production and consumption is creating an anti-landscape beyond the scope of recreational or residential redevelopment. The United Nations

Environmental Program announced in June 2012, that nations worldwide have made significant progress in only four of the ninety most important environmental challenges facing the world in the twenty-first century. "The report cautions that if humanity does not urgently change its ways, several critical thresholds may be exceeded, beyond which abrupt and generally irreversible changes to the life-support functions of the planet could occur" (UNEP 2012, 1). While clearly any restoration of scarred, unlivable land is better than nothing, transforming a former landfill into a pretty, urban park is not sufficient to address these much larger environmental problems.

Anti-landscapes, as a concept, direct environmental historians and other scholars to add elements of cultural and environmental sustainability to place theory. But if the anti-landscape is to be a useful construct, there is work to be done to integrate the kinds of case studies presented here into their broader historical and environmental contexts. The kinds of human activities that produce anti-landscapes, whether a sanitary landfill or a giant mining pit, grow out of decades of political, economic and social decisions. The evolution of the priorities and values that drive first the creation and then the restoration of anti-landscapes is a fascinating story in itself. This broader discussion also raises an interesting cultural riddle: at the margins of modern industrial production, which is the anti-landscape? Was Rio de Janeiro's gigantic Jardim Gramacho dump an anti-landscape when it supported a population of 5000 scavengers, or is it an anti-landscape now that it has closed? In the 1980s, were the dilapidated tenements of East Harlem, or the "counter space" of community gardens that sprouted in vacant lots among them more of an anti-landscape (Hayden 1995, 34-39)? When an oil boom brought a huge influx of population, industrial development and rampant real estate speculation to Los Angeles in the 1910s and 1920s, was this a landscape, by virtue of its community-building economic boom, or an anti-landscape because of the destruction wrought by oil wells, explosions and fires in the oil fields (Elkind 2011, 20-30)?

Bibliography

Booker, Matthew Morse. 2006. 'Oyster Growers and Oyster Pirates in San Francisco Bay' in *Pacific Historical Review* 75(1): 63-88.

Coleman, Annie Gilbert. 2004. *Ski Style: Sport and Culture in the Rockies*. Lawrence: University Press of Kansas.

Elkind, Sarah S. 2011. *How Local Politics Shape Public Policy: Business, Power and the Environment in Twentieth Century Los Angeles.* Chapel Hill: University of North Carolina Press.

Flores, Dan. 1994. 'An Argument for Bioregional History,' in *Environmental History Review* 18(4): 1-18.

Hayden, Dolores. 1995. *The Power of Place: Urban Landscapes as Public History.* Cambridge, Mass.: The MIT Press.

'Historien om Stige Ø'. On line at: http://odensefjord.dk/stigeoe/StigeoeHistorie/HistorieForside.html (consulted 10 June 2012).

Larsen, Rasmus S. 2010. *Historien om Stige Ø.* Odense: Odense Kommune.

Long, Patrick T. 2001. "For Residents and Visitors Alike: Seeking Tourism's Benefits, Minimizing Tourism's Costs," in Wrobel, David M. and Patrick T. Long (eds) *Seeing and Being Seen: Tourism in the American West.* Lawrence: University Press of Kansas: 70-99.

Nye, David. 2010. *When the Lights Went Out: A History of Blackouts in America.* Cambridge, Mass.: The MIT Press.

'Opsamling af forurenet regnvand'. On line at http://www.odenserenovation.dk/Odenserenovation/Stige_oe/Opsamling-af-forurenet-regnvand.aspx (consulted 12 June 2012).

Rothman, Hal. 1998. *Devil's Bargains: Tourism in the Twentieth-Century American West.* Lawrence: University Press of Kansas.

Thrush, Coll. 2007. *Native Seattle: Histories from the Crossing-Over Place.* Seattle: University of Washington Press.

United Nations Environmental Programme (UNEP).. 2012. 'World Remains on Unsustainable Track Despite Hundreds of Internationally Agreed Goals and Objectives,' 6 June 2012, on line at http://www.unep.org/newscentre/default.aspx?DocumentID=2688&ArticleID=9158 (consulted 7 June 2012).

Worster, Donald. 1979. *Dust Bowl: The Southern Plains in the 1930s.* New York: Oxford University Press.

Contributors

Hannes Bergthaller teaches at the Department of Foreign Languages and Literatures at National Chung-Hsing University in Taichung, Taiwan. His monograph *Populäre Ökologie* on the literature and cultural history of modern US environmentalism appeared in 2007; the collection *Addressing Modernity: Social Systems Theory and US Cultures* (co-edited with Carsten Schinko) was published in 2011. His primary research interests are ecocriticism, American literature, and social systems theory. He is currently president of the European Association for the Study of Literature, Culture and the Environment (EASLCE).

Werner Bigell teaches English at Finnmark University in Norway. He wrote his dissertation about Edward Abbey and is currently interested in questions of community and nature, the commons, ecomuseums, allotment and community gardens, as well as in class and cultural differences in the use and perception of natural spaces.

Sarah S. Elkind is a Professor of History teaching environmental, political and urban history at San Diego State University. She held the 2010-2011 Fulbright Distinguished Chair in American Studies at the University of Southern Denmark. She has published *How Local Politics Shape Federal Policy* (2011) on business influence in twentieth century American politics, and *Bay Cities and Water Politics* (1998) on regional public works and political reform in American cities. Before graduate school, she ran educational programs at parks reclaimed from a nineteenth-century military installation and an abandoned drive-in movie theater.

Maurits W. Ertsen is associate professor at Delft University of Technology, specializing in irrigation design and management in modern, historical and ancient settings. He has published on irrigation engineering in colonial regions such as Indonesia and Kenya. Furthermore, he works on agent-based models to build further understanding of everyday behavior and its longer term consequences in irrigation systems. He is editor of *Water History*, the journal of the International Water History Association.

James Rodger Fleming is professor of science, technology, and society at Colby College. Founder and first president of the International Commission on the History of Meteorology, he is editor of *History of Meteorology*, and series editor for *Palgrave Studies in the History of Science and Technology*. His books include *Meteorology in America, 1800-1870* (1990), *Historical Perspectives on Climate Change* (1998), *The Callendar Effect* (2007), and *Fixing the Sky* (2010). New work includes a book on Harry Wexler and the emergence of atmospheric science, a biography of the substance we now call carbon dioxide, and an exploration of the Anthropocene.

Anna Flügge teaches American Literature at Ludwig-Maximilians-University in Munich, Germany. Her fields of research include film adaptations and twentieth century poetry and fiction. The title of her dissertation is "James Ellroy and the Novel of Obsession." Currently, she is writing about wastelands in modern and contemporary American poetry and fiction.

Maunu Häyrynen is Professor at the School of Cultural Production and Landscape Studies, University of Turku. His PhD thesis pertained to the history of Helsinki public parks and park politics (1994). His research engages the construction of nationalistic landscape imagery, transboundary landscapes and landscape conservation. Publications include: "The Transboundary Landscape of the Schengen Border" (guest editor) *Journal of Borderlands Studies* 24.2 (2009); (together with Olavi Granö & Niina Käyhkö) "Finnish landscape studies – a mixture of traditions and recent trends in the analysis of nature- human interactions" in *BELGEO, Revue belge de géographie* 2.3 (2004); "A Kaleidoscopic Nation: The Finnish National Landscape Imagery" in *Nordic Landscapes, Region and Belonging on the Northern Edge of Europe*, eds. Jones & Olwig (2008); "Assessment" in Landscape Theory, eds. DeLue & Elkins (2008); and "Bildspråket i finländsk landskapsavbildning," in *Nationens röst, texter om nationalismens teori och praktik*, ed. Sverker Sörlin (2001).

Øyunn Hestetun is Associate Professor of American literature and culture at the Department of Foreign Languages, University of Bergen, Norway. She received her doctorate in American literature from Uppsala University, Sweden, in 1993. Her main area of research is literary and cultural theory, and her most recent publications focus on postcolonial theory, Native American literature, migrant writing, and literature and the environment.

Nikolet Jensen works as a project coordinator and online media editor for the Credo Bonum Foundation in Sofia, focusing on culture and sustainability, and she also publishes from time to time in Bulgarian print media. She received a BA in journalism from the University of Oklahoma and worked for 4 years in New York at Brew Media Relations, before taking her MA in American Studies at the University of Southern Denmark.

Mark Luccarelli is Associate Professor in the Department of Literature, Area Studies and European Languages, University of Oslo. He is one of the founders and organizers of the Nordic Network for Interdisciplinary Studies (NIES), co-editor (with P. G. Roe) of *Green Oslo: Visions, Planning and Discourse* (2013) and author of *Lewis Mumford and the Ecological Region* (1995) and several articles on space, landscape, and representations of nature.

David E. Nye is professor and founder of the Center for American Studies at the University of Southern Denmark. Author of more than 150 publications, he received the two book prizes (1993, 2009) given by the Society for the History of Technology and the Leonardo da Vinci Medal (2005). Recent work includes *Technology Matters* (2006), *When the Lights Went Out* (2010) and *America's Assembly Line* (2013), all with MIT Press. He has been a visiting professor at Cambridge, MIT, Harvard, Warwick, and Notre Dame. He is one of the founders of the Nordic Network for Interdisciplinary Studies (NIES).

Anna Storm is a postdoctoral researcher at the Centre for Baltic and East European Studies (CBEES), Södertörn University, Stockholm, Sweden. Her doctoral thesis was *Hope and Rust: Reinterpreting the industrial place in the late twentieth century* (Royal Institute of Technology, Stockholm 2008) She has received the Joan Cahalin Robinson Prize (2006), the ICOHTEC Prize for Young Scholars (2009) and the Marie Nisser scholarship (2012). Currently she holds a postdoctoral position funded by the Swedish Research Council with an individual research project on “Politics of Remembering: Contested heritage processes at Ignalina and Barsebäck nuclear power plants.”

Index

www.ingramcontent.com/pod-product-compliance
Lightning Source LLC
Chambersburg PA
CBHW071443230926
51846CB00001B/324
9789042038868